# Neuropeptide Y Protocols

# METHODS IN MOLECULAR BIOLOGY™

METHODS IN MOLECULAR BIOLOGY™

# Neuropeptide Y
# Protocols

Edited by

## Ambikaipakan Balasubramaniam

*Department of Surgery*
*University of Cincinnati Medical Center*
*Cincinnati, OH*

Humana Press ✳ Totowa, New Jersey

This publication is printed on acid-free paper. ∞

ANSI Z39.48-1984 (American Standards Institute) Permanence of Paper for Printed Library Materials.

Cover design by Patricia F. Cleary.

For additional copies, pricing for bulk purchases, and/or information about other Humana titles, contact Humana at the above address or at any of the following numbers: Tel: 973-256-1699; Fax: 973-256-8341; E-mail: humana@humanapr.com, or visit our Website at www.humanapress.com

**Photocopy Authorization Policy:**

Printed in the United States of America. 10 9 8 7 6 5 4 3 2 1

Library of Congress Catologing-in-Publication Data

Neuropeptide Y protocols/edited by Ambikaipakan Balasubramaniam.
       p. cm.--(Methods in molecular biology;v. 153)
    Includes bibliographical references and index.
    ISBN 0-89603-662-6 (alk paper)
    1. Neuropeptide Y--Laboratory manuals. I. Balasubramaniam,
Ambikaipakan. II. Methods in molecular biology (Totowa, N.J.); v. 153.
    QP552.N38 N476 2000
    573.8'465--dc21
                                        00-020392

# Preface

The observation that neuropeptide Y (NPY) is the most abundant peptide present in the mammalian nervous system and the finding that it elicits the most powerful orexigenic signal have led to active investigations of the properties of the NPY family of hormones, including peptide YY (PYY) and pancreatic polypeptide (PP). Nearly two decades of research have led to the identification of several NPY receptor subtypes and the development of useful receptor selective ligands. Moreover, these investigations have implicated NPY in the pathophysiology of a number of diseases, including feeding disorders, seizures, memory loss, anxiety, depression, and heart failure. Vigorous efforts are therefore continuing, not only to understand the biochemical aspects of NPY actions, but also toward developing NPY-based treatments for a variety of disorders. To facilitate these efforts, it was decided to produce the first handbook on NPY research techniques as part of the Methods in Molecular Biology Series.

In compiling *Neuropeptide Y Protocols*, I have gathered contributions on techniques considered critical for the advancement of the NPY field from experts in various disciplines. Each chapter starts with a brief introduction, with Materials and Methods sections following. The latter sections are presented in an easy to follow step-by-step format. The last section of the chapter, Notes, highlights pitfalls and the maneuvers employed to overcome them. This information, not usually disseminated in standard research publications, may prove extremely useful for investigators employing these techniques in NYP research.

*Neuropeptide Y Protocols* contains a total of 18 chapters divided into five parts. Part I describes various cloning techniques in six chapters, including genomic DNA isolation, expression cloning, classical techniques, PCR cloning, construction of hybrid receptors, and homology-based cloning. Production of transgenic and knockout models is described in Part II. Four chapters in Part III illustrate the use of antisense technology to define the receptors and the signal transduction pathways mediating NPY actions. Various qualitative and quantitative techniques used to study tissue mRNA distribution are described in a total of five chapters spanned across Parts IV and V. A chapter on radioligand binding is also included in Part V. The techniques described here could easily be extrapolated to study any peptide hormone. Therefore, *Neuropeptide Y Protocols* should benefit all investigators involved in polypeptide hormone research.

I wish to express my appreciation to all the authors for their excellent contributions, and particularly for meeting their deadlines. Appreciation is also expressed to my colleague, Sulaiman Sheriff, for the chapter and his help in selecting the contents of this volume. The series editor, John Walker, has contributed to the success of this volume in many ways, and is appreciated very much. Finally, I am grateful to Koti Sreekrishna, Senior Scientist, Procter and Gamble, Inc., Cincinnati, OH, for his expert help in editing these chapters.

### *Ambikaipakan Balasubramaniam*

# Contents

# Contributors

DAVID A. AMMAR • *Department of Molecular and Cell Biology, University of California, Berkeley, CA*

AMBIKAIPAKAN BALASUBRAMANIAM • *Department of Surgery, University of Cincinnati Medical Center, Cincinnati, OH*

MAGNUS M. BERGLUND • *Department of Neuroscience and Pharmacology, Uppsala University, Uppsala, Sweden*

WILLIAM T. CHANCE • *Department of Surgery, University of Cincinnati Medical Center, Cincinnati, OH*

CHAD FORADORI • *Department of Psychiatry, Kettering Laboratory, University of Cincinnati School of Medicine, Cincinnati, OH*

MATHIEU GOUMAIN • *INSERM U.410, Paris, France*

HERBERT HERZOG • *Garvan Institute of Medical Research, St. Vincent's Hospital, Darlinghurst, Sydney, Australia*

KARL G. HOFBAUER • *Metabolic and Cardiovascular Research, Novartis Pharma AG, Basel, Switzerland*

PUSHPA S. KALRA • *Department of Physiology, University of Florida Brain Institute, University of Florida College of Medicine, Gainesville, FL*

SATYA P. KALRA • *Department of Neurosciences, University of Florida Brain Institute, Gainesville, FL*

PETER KRISTENSEN • *Departments of Histology and Health Care Discovery, Novo Nordisk, Bagsvard, Denmark*

MARC LABURTHE • *INSERM U.410, Paris, France*

JENNIFER LACHEY • *Department of Psychiatry, Kettering Laboratory, University of Cincinnati School of Medicine, Cincinnati, OH*

DAN LARHAMMAR • *Department of Neuroscience and Pharmacology, Uppsala University, Uppsala, Sweden*

PHILIP J. LARSEN • *Department of Anatomy, The Panum Institute, Copenhagen, Denmark*

INGRID LUNDELL • *Department of Neuroscience and Pharmacology, Uppsala University, Uppsala, Sweden*

ANNE-MARIE LORINET, *INSERM U.410, Paris, France*

DOUGLAS J. MACNEIL • *Merck and Co., Inc., Rahway, NJ*

KAREN E. MCCREA • *Garvan Institute of Medical Research, St. Vincent's Hospital, Darlinghurst, Sydney, Australia*

MIECZYSLAW MICHALKIEWICZ • *Department of Physiology, Health Science Center, West Virginia University, Morgantown, WV*

TERESA MICHALKIEWICZ • *Department of Physiology, Health Science Center, West Virginia University, Morgantown, WV*

ERIC M. PARKER • *Department of CNS and Cardiovascular Research, Schering-Plough Research Institute, Kenilworth, NJ*

RACHEL PARKER • *Garvan Institute of Medical Research, St. Vincent's Hospital, Sydney, Australia*

THIERRY PEDRAZZINI • *Division of Hypertension, University of Lausanne Medical School, Lausanne, Switzerland*

JENNIFER PROFITT • *Department of Psychiatry, Kettering Laboratory, University of Cincinnati School of Medicine, Cincinnati, OH*

ABHIRAM SAHU • *Department of Cell Biology and Physiology, University of Pittsburgh School of Medicine, Biomedical Science Tower, Pittsburgh, PA*

ANDREA O. SCHAFFHAUSER • *Lonza Ltd., Basel, Switzerland*

RANDY J. SEELEY • *Department of Psychiatry, Kettering Laboratory, University of Cincinnati School of Medicine, Cincinnati, OH*

JOSIANE SEYDOUX • *Department of Physiology, University of Geneva, Geneva, Switzerland*

SULAIMAN SHERIFF • *Department of Surgery, University of Cincinnati Medical Center, Cincinnati, OH*

ALAIN STRICKER-KRONGRAD • *Metabolic Diseases, Millennium Pharmaceuticals, Cambridge, MA*

DEBRA A. THOMPSON • *Department of Ophthalmology and Visual Sciences and Department of Biological Chemistry, University of Michigan, Ann Arbor, MI*

THIERRY VOISIN • *INSERM U.410, Paris, France*

DAVID H. WEINBERG • *Merck and Co., Inc., Rahway, NJ*

CAROLYN A. WORBY • *Department of Biological Chemistry, University of Michigan, Ann Arbor, MI*

# I

## CLONING TECHNIQUES

# 1

# Cloning Neuropeptide Tyrosine cDNA

## Carolyn A. Worby

## 1. Introduction

The nervous system is composed of a diverse population of cells selectively expressing different genes which ultimately determine the type of neurotransmitters and neuropeptides produced and released. Neuropeptide tyrosine (NPY) is one of the most abundant and widely expressed peptides in the mammalian nervous system (*1*). It was originally isolated by virtue of a chemical assay that detects the presence of a carboxyl-terminal amide in proteins (*2*); a hallmark of potential hormonal function. Its widespread expression in the central and peripheral nervous systems, in addition to the carboxyl-terminal amide, suggested that NPY could be a critical neurotransmitter in a variety of neuronal processes. We were therefore interested in obtaining the cDNA for human NPY in order to study NPY gene expression and screen for the NPY gene (*3,4*). The only available information was the amino acid sequence of porcine NPY and the amino acid sequences of two related peptides, pancreatic polypeptide and peptide YY (*5,6*). These experiments were performed in the early 1980s before the advent of genome sequencing and polymerase chain reaction (PCR) technology. Today, experiments to obtain cDNAs for proteins for which only amino acid sequence is available are designed very differently and include computer searches of nucleotide databases and/or various PCR strategies. The experiments described in this chapter represent the classical way by which cDNAs were cloned from corresponding protein sequence.

## 2. Materials

Avian myeloblastosis virus (AMV) reverse transcriptase was provided by J. Beard (Life Sciences, St. Petersburg, FL). RNasin was purchased from Promega (Madison, WI). DNA polymerase I was purchased from Boehringer

From: *Methods in Molecular Biology, vol. 153: Neuropeptide Y Protocols*
Edited by: A. Balasubramaniam © Humana Press Inc., Totowa, NJ

Mannheim (Indianapolis, IN). γ-Labeled nucleotide triphosphates and α-labeled nucleotide triphosphates were purchased from Amersham/Pharmacia (Piscataway, NJ). Terminal deoxynucleotidyltransferase (TDT) and S1 nuclease (from *Aspergillus oryzae*) were purchased from Gibco/BRL (Gaithersburg, MD).

1. Guanidinium isothiocyanate (Gu-HSCN) extraction buffer: 4 $M$ Gu-HSCN, 1% Sarkosyl, 50 m$M$ EDTA, 25 m$M$ lithium citrate, pH 7.0. Add just before use: 0.1% antifoam A and 0.1 $M$ β-mercaptoethanol.
2. CsCl solution: 5.7 $M$ CsCl, 0.1 $M$ EDTA.
3. TEC buffer: 10 m$M$ triethylammonium carbonate, pH 7.4.
4. S1 buffer: 3 $M$ sodium acetate, pH 4.5, 3 $M$ NaCl, 45 m$M$ zinc chloride.
5. TNE buffer: 10 m$M$ Tris-HCl, pH 7.5, 50 m$M$ NaCl, 1 m$M$ EDTA.
6. Tailing buffer: 1.4 $M$ potassium cacodylate, 1 m$M$ 1,4-dithiothreitol (DTT), 100 m$M$ CoCl$_2$.
7. Annealing buffer: 10 m$M$ Tris-HCl, pH 7.4, 100 m$M$ NaCl, 1 m$M$ EDTA.
8. 2XYT: 1.6% tryptone, 1.0% yeast extract, 0.5% NaCl.
9. Kinase denaturation buffer: 5 m$M$ Tris-HCl, pH 9.5, 1 m$M$ Spermidine, 0.1 m$M$ EDTA.
10. Kinase buffer: 500 m$M$ glycine, pH 9.5, 100 m$M$ MgCl$_2$, 50 m$M$ DTT, 50% glycerol.
11. 1X NaCl/Cit: 0.15 $M$ NaCl, 0.015 $M$ sodium citrate.
12. 1X Denhardt's solution: 0.02% bovine serum albumin, 0.02% Ficoll, 0.02% polyvinylpyrrolidone.

## 3. Methods

### 3.1. Library Construction

#### 3.1.1. Isolation of Total RNA

Total RNA is isolated by the Gu-HSCN method of Chirgwin et al. *(7)* *(see* **Note 1**). The CsCl solution is treated with 0.1% diethylpyrocarbonate (DEPC) by stirring for 1–2 h followed by autoclaving for 40 min to remove the DEPC. All solutions (except the Gu-HSCN extraction buffer) that come into contact with the purified RNA should be similarly treated. Glassware is baked at 300°C overnight. These precautions are taken to ensure that no RNase is present.

1. Homogenize 1 g of tissue per 20 mL Gu-HSCN extraction buffer for 30–60 s using a Polytron (Kinematica, Sweden) at half speed.
2. Centrifuge homogenate for 10 min at 8000$g$. Pellet should be small.
3. Layer Gu-HSCN solution over 1.2 mL CsCl solution in cellulose nitrate SW 50.1 tubes. (Beckman Instruments, Fullerton, CA).
4. Centrifuge in SW 50.1 rotor at 20°C for 17 h at 36,000 rpm (120,000$g$).
5. Remove and discard supernatant. RNA will be a clear pellet.
6. Resuspend the RNA in the Gu-HSCN solution without antifoam A. A Dounce homogenizer may aid in the suspension of the pellet.
7. Centrifuge as in **step 2**. Discard pellet, which should be very small.

8. Precipitate supernatant with 0.1 volume of 3 *M* potassium acetate, pH 5.0 (DEPC - treated), and 2 volumes of ethanol. Incubate at –20°C for at least 2 h.
9. Centrifuge precipitated RNA as in **step 2**.
10. Rinse pellet with 70% ethanol. Briefly air dry. Do not allow pellet to dry completely.
11. Dissolve pellet in 10 m*M* Tris-HCl, pH 7.5. Dounce homogenize and/or heat to 65°C for 5–10 min if necessary to solubilize pellet.
12. Centrifuge as in **step 2**. Discard pellet.
13. Precipitate supernatant as in **step 8**.
14. Redissolve pellet in DEPC-treated water. No homogenization should be necessary.
15. Store precipitated RNA at –20°C or dissolve in water and store at –80°C. Read $A_{260}$ for determination of concentration (42 µg RNA/1.0 $A_{260}$).

## 3.1.2. Construction of cDNA Clones

Typically, the total RNA prepared in the previous section would be further purified by oligo (dT)-cellulose chromatography in order to obtain poly (A) RNA *(8)* (*see* **Note 2**). However, in this case, the human tissue sample was small and yielded only a few hundred micrograms of total RNA. This was not enough to subject to oligo (dT)-cellulose chromatography. Therefore, total RNA served as the template for ds-cDNA synthesis using the following reaction.

### 3.1.2.1. SYNTHESIS OF DS-cDNA

*First strand synthesis:*

> 30 µg total RNA [approx 0.6 µg of poly (A) RNA].
> 2 µL 10X buffer (supplied by manufacturer of reverse transcriptase).
> 2 µL 10 m*M* dNTPs (each nucleotide is 10 m*M*).
> 2 µL 300 m*M* β-mercaptoethanol.
> 2 µL 1 µg/µL oligo dT.
> 1 µL RNasin.
> 1 µL AMV reverse transcriptase.
> DEPC'd $H_2O$ to 20 µL.

Incubate at 42°C for 1 h.

*Second strand synthesis:*

Heat first strand reaction for 3 min at 90°C, quick cool on ice, and centrifuge at 16,000*g* for 5 min. This step pellets the denatured AMV reverse transcriptase. Remove supernatant to a new tube. Add:

> 19 µL first strand reaction.
> 4 µL 1 *M* HEPES pH 6.9.
> 1 µL 10 m*M* dNTPs.
> 10 U DNA polymerase I.
> $H_2O$ to 40 µL.

Incubate for 3 h at 15°C. Add NaOH to a final concentration of 0.3 *M*, and incubate for 20 min at 46°C to hydrolyze the RNA. Readjust the pH to approx 7 with 1 *M* HCl. Pass the reaction mixture over a Sephadex G-100 column (1 × 10 cm) equilibrated in 10 m*M* triethylammonium carbonate (TEC), pH 7.4. Pool the samples containing the cDNA and vacuum dry. TEC is a good buffer of choice for this column. Because it is volatile, the sample can be recovered by vacuum drying with very little salt residue. At this step, the ds-cDNA could also be concentrated by ethanol precipitation using 0.1 volume of 3 *M* sodium acetate, pH 5.0 and 3 volumes of ethanol. All subsequent DNA precipitation steps are done in this manner.

### 3.1.2.2. S1 TREATMENT

Resuspend the dried cDNA in $H_2O$ and digest with S1 nuclease according to the following reaction:

150 µL ds-cDNA.
4 µL S1 nuclease (0.5 U/mL).
17.5 µL 10 x S1 buffer.
$H_2O$ to 175 µL.

Incubate 1 h at 37°C. Terminate the reaction by addition of 10 µL of 0.5 *M* EDTA followed by chloroform/isoamyl alcohol (24:1) extraction and ethanol precipitation (*see* **Note 3**).

### 3.1.2.3 SIZE SELECTION OF THE DS-CDNA

1. Place the S1-trimmed ds-cDNA onto a 5–20% linear sucrose gradient (10 mL) and centrifuge at 175,000*g* at 15°C for 16–18 h.
2. Collect fractions (0.5 mL). The ds-cDNA can be followed either by scintillation counting (if radiolabeled nucleotides are used) or $A_{260}$ determination.
3. Collect the fractions containing the higher molecular weight ds-cDNAs and precipitate with ethanol.

If the ds-cDNA is radiolabeled, its size can be determined by running a few microliters of each fraction on a 1% agarose gel followed by autoradiography. The sucrose gradients are made by placing 1.7 mL of 200 mg/mL sucrose in the bottom of an SW41 nitrocellulose tube (Beckman Instruments) followed by 3.4 mL of 150 mg/mL sucrose, 3.4 mL of 100 mg/mL sucrose, and 1.7 mL of 50 mg/mL sucrose. All sucrose solutions are made in TNE buffer. The sucrose solutions are carefully layered and then allowed to diffuse for 24 h at 4°C (*see* **Note 4**).

### 3.1.2.4. TAILING AND ANNEALING OF THE DS-CDNA AND VECTOR (PUC8)

Tailing of the ds-cDNA or vector is carried out using the following reaction mix:

50 µL ds-cDNA or *Pst*I-digested pUC8 (X pM of 3'-hydroxyl groups).
6 µL 10X tailing buffer.
1 µL 10 m*M* dCTP (cDNA) or 10 m*M* dGTP (pUC8).
TDT (8.5 U/pmol of 3'-hydroxyl groups for the cDNA or 17 U of TDT/pmol of
 3'-hydroxyl groups for the vector).
H$_2$O to 60 µL.

1. Incubate the above reaction for 4 min at 37°C.
2. Remove the unincorporated nucleotides by applying these reaction mixtures to a Sephadex G-100 column. Recover the cDNA or vector as previously described (*see* **Subheading 3.1.2.1.**).
3. Anneal the tailed plasmid and ds-cDNA at a 1:1 molar ratio in annealing buffer. The final total DNA concentration should be 3 µg/mL.
4. Heat this mixture for 3 min at 65°C and 2 h at 42°C, and then cool gradually to room temperature (*see* **Note 5**).

### 3.1.3. Transformation

*Escherichia coli* strain JM83 is made competent for transformation by calcium shock as described **(9)**. This technique will not be described in detail because very high efficiency competent cells can be purchased from a variety of sources.

1. Competent cells are transformed with the annealed mixture by incubating for 5 min at 4°C, 3 min at 42°C, and 10 min at room temperature.
2. These cells are then diluted into 50 volumes of bacterial media (2XYT) and recovered for 2 h with shaking at 37°C.
3. The cells are pelleted by centrifugation at 5000*g* followed by plating on the above media containing 1.2% agar and 50 µg/µL ampicillin.

If 5-bromo-4-chloro-3-indolyl-β-D-galactosidase (X-gal) is added to the agar, the bacterial colonies that contain the recombinant vector will be white, while those containing vector without an insert will be blue **(9)**. This is a good method for determining the complexity of your library (*see* **Note 6**).

## 3.2. Screening the cDNA Library

### 3.2.1 Hybridization Probes

Transformants containing the NPY cDNA are identified by using the mixed hybridization probe 5'-A-(A,G)-(A,G)-T-T-(A,G,T)-A-T-(A,G)-T-A-(A,G)-T-G-3' to screen the library. This sequence is derived from the known porcine amino acid sequence, His-Tyr-Ile-Asn-Leu. The strategy for design of the probe is twofold. First, the portion of the protein that is the most highly conserved among pancreatic polypeptide, peptide YY, and NPY is evaluated. We determined that the carboxyl terminus is more highly conserved between family members and therefore more likely to be highly conserved between the porcine and human NPY sequences. Second, we want to choose amino acids that have the fewest codon choices so that the degeneracy of our probe will be minimized. This probe is 5'-end-labeled by polynucleotide kinase to a specific activity of approximately $1 \times 10^8$ cpm/µg. The following reaction protocol is employed:

1. Resuspend 100 pmol of oligonucleotide in 22 µL of kinase denaturation buffer.
2. Heat for 7 min at 90°C, quick cool on ice, and centrifuge for 2–3 min at 16,000$g$.
3. Remove 20 µL of supernatant, taking care not to transfer any pelleted material. Add to the following:
      5 µL of [γ-$^{32}$P] ATP (dried).
      3 µL kinase buffer.
      1 µL T4 polynucleotide kinase.
   Incubate for 30 min at 37°C.
4. Separate free label from labeled probe by Sephadex G-50 chromatography (1 × 30 cm). This is an important step as free label will contribute significantly to the background of the colony hybridization (below).

### 3.2.2 Colony Hybridization

Transformants are screened using the $^{32}$P-labeled tetradecamers by the method of Grunstein and Hogness *(10)*.

1. Lift the bacterial colonies onto nitrocellulose and place colony side up on sheets of Whatman 3 MM paper saturated with 0.5 $M$ NaOH for 5 min.
2. The filters are then neutralized on 3 MM paper saturated with 1 $M$ Tris-HCl, pH 7.4, for 5 min.
3. Repeat **step 2**.
4. Place the filters on 3 MM paper saturated with 0.5 $M$ Tris-HCl, pH 7.4, 1.5 $M$ NaCl for 5 min.
5. Dry the filters in a vacuum oven at 80°C between Whatman paper for 2 h.
6. Prehybridize the filters for 4 h at 37°C in 5X NaCl/Cit, 10X Denhardt's solution, and *E. coli* DNA (500 µg/mL).
7. Hybridize the filters for 30 h at room temperature in the above buffer including 300 mg of Torula yeast RNA and $2 \times 10^5$ cpm of probe/mL.
8. Wash the filters twice on ice for 15 min in 4X NaCl/Cit/1x Denhardt's, once at room temperature for 15 min in 5X NaCl/Cit/0.2% SDS, and twice at 30°C for 15 min in 5X NaCl/Cit/0.2% SDS.
9. Positive colonies are visualized by autoradiography (*see* **Note 7**).

## 4. Notes

1. Isolation of RNA by this method requires the preparation of RNAse free solutions and glassware. Always wear gloves during the preparation of these materials as well as during the procedure itself. This technique works very well but

requires the availability of an ultracentrifuge and rotor as well as being more time consuming than newer protocols using TRIzol (Gibco/BRL, Gaithersburg, MD). If faced with the task of making RNA from several grams of tissue, I would choose the Chirgwin protocol. If I had a very small amount of tissue or tissue culture cells from which to isolate RNA, the TRIzol method would be preferable. Finally the biggest problem one faces when making RNA from tissues is the handling of the tissue. Tissue samples should be collected as quickly as possible and frozen immediately in small easily homogenized pieces on dry ice or in liquid nitrogen. For long-term storage, the tissue should be frozen either at –80°C or in liquid nitrogen. Once frozen, tissue can be broken with a hammer by placing it between two weighing boats filled with dry ice. It is my experience that the quality of your RNA depends more on how the tissue was collected then the sterility of your reagents. Finally, it is of critical and obvious importance to use a tissue source for RNA isolation that expresses the protein you are trying to clone.

2. It is a good idea to isolate poly(A) RNA for use in cDNA preparation. This may not always be possible depending on the amount of starting material. Typically, poly(A) RNA amounts to 2% of the total RNA. Therefore, it is preferable to have at least 1 mg of total RNA before attempting to isolate poly(A) RNA using conventional oligo (dT)-cellulose chromatography. However, it is possible to isolate poly(A) RNA from smaller amounts of total RNA using kits supplied by various manufacturers, for example, the PolyATtract kit (Promega, Madison, WI). The first strand reaction usually contains 1 µg of poly (A) RNA/10 µL of reaction volume. The reaction volume is then doubled for the second strand reaction. It is also a good idea to add a trace amount of radioactive nucleotide to your first and second strand synthesis reactions in order to quantitate the amount of ds-cDNA that is synthesized. For example, trace amounts of $[^3H]dCTP$ can be used to label the first strand synthesis reaction. After completion of the reaction, the percent incorporation of the radiolabeled nucleotide is calculated by subjecting a microliter of the reaction to column chromatography using the Sephadex G-100 column described above. The calculation is performed as described below:

*Calculation of the amount of cDNA synthesized:*

1 m$M$ nucleotides (incorporation) (4 nucleotides)(reaction volume) = mmol
mmol (327 mg/mmol) = mg cDNA synthesized

where incorporation = total cpm of first peak (i.e., cDNA)/cpm of the last peak (i.e., free nucleotides) and 327 is an estimate of the molecular weight of a nucleotide.

*Calculation of the % first strand synthesis:*

% first strand synthesis = (µg of first strand/µg of starting poly(A) RNA) × 100%

The percent first strand synthesis should be 40% or greater. This value depends on the quality of the poly(A) RNA. Ribosomal RNA will lower the percent first strand synthesis. If your RNA is contaminated with RNAse, percent first strand synthesis will be very poor (<5%).

The amount of second strand synthesized can be calculated in a similar manner by adding a small amount of $^{32}$P-labeled nucleotide such as [$\alpha$-$^{32}$P]dCTP to the reaction. The percent second strand synthesis should approach 100%. This $^{32}$P-labeled ds-cDNA can be more easily followed in the subsequent reactions and will serve as a means of calculating the amount of cDNA remaining after subsequent steps. Because decay of the $^{32}$P-labeled nucleotide can break the cDNA strand, it is a good idea to proceed directly from this step to the transformation of the bacteria as quickly as possible. The generation of the library should take no longer than 5 days.

3. This reaction removes the hairpin loops generated by AMV reverse transcriptase and creates blunt ends on the ds-cDNA. In some cases it is necessary to use Klenow fragment to fill in the ends *(9)*. This step is more important if linkers are to be added; the tailing reaction described later is generally effective even if the ends of the cDNA are not perfectly blunted.

4. This step is especially important when using oligo (dT) as the primer for first strand synthesis. If the cDNAs are not size selected, the library may consist mostly of very short cDNAs containing only the 3'-termini of the mRNAs. When precipitating the cDNAs from the fractions, it may be necessary to use an ultracentrifuge (50,000$g$, 30 min at 4°C). At this point, you may have very little cDNA in several milliliters of solution. It is also possible to use glycogen (5–10 mg) as a carrier in the precipitation step to avoid losses.

5. The timing of this reaction is critical for efficient tailing of the cDNA and vector. *Pst*I-digested pUC8 is tailed with dGTP and trace amounts of [$^3$H]dGTP for varying lengths of time. The number of dG residues added per 3'-hydroxyl is calculated. The reaction time that results in the addition of 15 dG residues per end is used. Since dC is added at a different rate than dG, a similar set of experiments must be done for the C-tailing reaction. It is very important to digest the vector to completion prior to tailing with dGTP. If trace amounts of supercoiled vector remain, it will transform the bacteria with a much higher efficiency than the annealed cDNA/vector, resulting in a large number of background colonies that contain only vector.

6. It is better to buy the high efficiency competent cells than to make them yourself, since transformation efficiency obtained with the commercial cells is far superior to that obtained for cells made competent by traditional calcium shock methods. Commercial competent cells will come with a suggested transformation protocol that should be followed as described.

7. The baking step is commonly replaced by ultraviolet (UV) crosslinking *(9)*. After the filters are baked or UV crosslinked, it is important to remove excess colony debris. This is accomplished by wetting the filters in prehybridization buffer and rubbing the colony debris from the filter using gloved hands. The cleaned filters should then be placed in fresh prehybridization buffer. When screening with a short degenerate oligonucleotide, it is easy to wash the filters too stringently. It is a good idea to monitor the level of radioactivity on the filters with a Geiger counter and expose the filters when they retain a modest level of radioactivity. If

the filters are not allowed to dry during exposure, i.e., if they are sandwiched between layers of Saran Wrap, they can be rewashed at a higher stringency if the background is unacceptable.

## References

1. Adrian, T. E., Allen, J. M., Bloom, S. R., Ghatei, M. A., Rossor, M. N., Roberts, G. W., et al. (1983) Neuropeptide Y distribution in human brain. *Nature* **306,** 584–586.
2. Tatemoto, K., Carlquist, M., and Mutt, V. (1982) Neuropeptide Y—a novel brain peptide with structural similarities to peptide YY and pancreatic polypeptide. *Nature* **296,** 659–660.
3. Minth, C. D., Bloom, S. R., Polak, J. M., and Dixon, J. E. (1984) Cloning, characterization, and DNA sequence of a human cDNA encoding neuropeptide tyrosine. *Proc. Natl. Acad. Sci. USA* **81,** 4577–4581.
4. Minth, C. D., Andrews, P. C., and Dixon, J. E. (1986) Characterization, sequence, and expression of the cloned human neuropeptide Y gene. *J. Biol. Chem.* **261,** 11,974–11,979.
5. Leiter, A. B., Keutmann, H. T., and Goodman, R. H. (1984) Structure of a precursor to human pancreatic polypeptide. *J. Biol. Chem.* **259,** 14,702–14,705.
6. Leiter, A. B., Toder, A., Wolfe, H. J., Taylor, I. L., Cooperman, S., Mandel, G., et al. (1987) Peptide YY. Structure of the precursor and expression in exocrine pancreas. *J. Biol. Chem.* **262,** 12,984–12,988.
7. Chirgwin, J. M., Przybyla, A. E., MacDonald, R. J., and Rutter, W. J. (1979) Isolation of biologically active ribonucleic acid from sources enriched in ribonuclease. *Biochemistry* **18,** 5294–5299.
8. Aviv, H. and Leder, P. (1972) Purification of biologically active globin messenger RNA by chromatography on oligothymidylic acid-cellulose. *Proc. Natl. Acad. Sci. USA* **69,** 1408–1412.
9. Ausubel, F. M., Brent, R., Kingston, R. E., Moore, D. D., Seidman, J. G., Smith, J. A., et al., eds. (1987) *Current Protocols in Molecular Biology.* Greene Wiley-Interscience, New York.
10. Grunstein, M. and Hogness, D. S. (1975) Colony hybridization: a method for the isolation of cloned DNAs that contain a specific gene. *Proc. Natl. Acad. Sci. USA* **72,** 3961–3965.

# 2

## Human Y1/Y5 Receptor Gene Cluster

*Isolation and Characterization*

### Herbert Herzog

### 1. Introduction

The methods used to isolate genomic clones of a particular gene have changed significantly over the years. In particular, the characterization of genes with many exons, separated by large introns, increased the need for clones with large inserts. The average length of an insert contained within one of the classical used λ cloning vectors (λgt10 or EMBL3) ranges from 10 to 15 kb. This insert size might be large enough to contain entire genes that contain no or only a few introns, but it is likely that such clones harbor only parts of larger genes. In many cases several overlapping λ clones must be identified and isolated by repeated screening ("chromosomal walking") of a library with end fragments isolated from the first clone. This significantly increases the time and effort involved in isolating a gene. In addition, the isolation of λ DNA is a laborious and time-consuming exercise that yields only low amounts of DNA.

Genomic libraries constructed in cosmid vectors have greatly improved the isolation of genes (*1,2*). This is mainly because of their capacity for harboring larger inserts of up to 40 kb and also because DNA can be conveniently isolated as a plasmid. However, to characterize and map very large genes or gene clusters it is desirable to have even larger inserts.

P1 and PAC/BAC vectors fulfil that criterion (*3,4*). Such clones contain high-molecular-weight inserts (75–100 or 120+ kb), about four to six times larger than λ clones, and two to three times larger than cosmids. In addition, the low copy number of the P1, PAC, or BAC vector (together with culture conditions such as growing them in restriction- and recombination-deficient *Escherichia coli* hosts), provides strongly improved stability of these clones.

From: *Methods in Molecular Biology, vol. 153: Neuropeptide Y Protocols*
Edited by: A. Balasubramaniam © Humana Press Inc., Totowa, NJ

Also, the isolation of super-coiled plasmids still follows standard plasmid purification protocols. Microgram quantities of plasmid DNA can be generated without any host genome contamination, which is always a problem with yeast artificial chromosomes (YACs), the only method available for cloning extremely large DNA (>1 million bp).

A further advantage of P1 and PAC/BAC clones is the fact that a relative low number of clones is required to cover a entire genome. For example, $10^5$ clones with an average insert size of 100 kb theoretically represent three times the human genome. This is normally a large enough number of independent clones to identify a positive signal successfully. This relative low number of clones is amenable to semiautomated screening procedures. Master filter sets on which individual P1 or PAC/BAC clones are spotted at a very high density can be generated, allowing the quick screening of only a few filters. Each DNA spot on the filter corresponds to a location on a microtiter plate that contains the library. Clones from putative hybridization signals can then be recovered from these microtiter plates and used for further analysis.

An additional advantage of this method is the direct identification of a single clone without going through second and third screening rounds. Positive signals can also easily be distinguished from false hybridization signals, because every clone is spotted twice in a special pattern within each small square so that true-positive clones always give two signals. The filters can also be screened many times (we have successfully used one filter set 18 times); this saves money and avoids repetitive plating of the library, which is often accompanied by an inherited decrease in titer.

A wide range of very different ligands (including neurotransmitters, neuropeptides, polypeptide hormones, and other bioactive molecules) transduce their signal to the intracellular environment by specific interaction with a class of receptors that relies on interaction with intracellular guanosine-5'-triphosphate (GTP)-binding protein (G-proteins). Molecular cloning has led to the identification of several hundred discrete G-protein-coupled receptors, and it has been demonstrated that around 80% of known hormones and neurotransmitters, including the neuropeptide Y (NPY) family, mediate their signal by activating G-protein-coupled receptors *(5)*. This makes the G-protein-coupled receptor superfamily, with its common structural and functional features, the largest single class of eukaryotic receptor, having an estimated 2000–3000 individual members. The structure of families of receptors exemplified by those coupled to G-proteins shows they are products that have evolved from an ancestral gene by duplication *(6)*. Many of these receptor genes lack introns, supporting the proposition that they were created via RNA-mediated transpositional events *(7)*. However, other mechanism of gene amplification may have occurred as well.

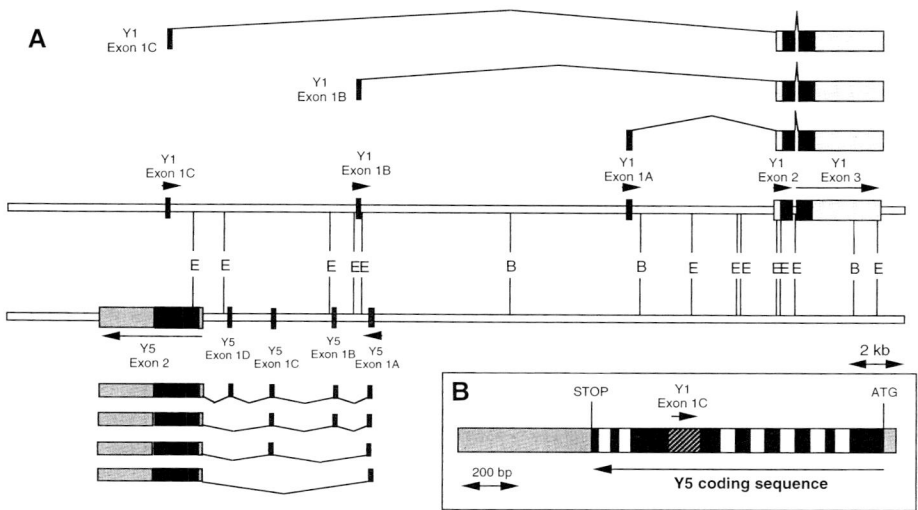

Fig. 1. The human Y1 and Y5 gene cluster. **(A)** A complete restriction map for the enzymes BamHI and EcoRI of the 30-kb region containing the genes for Y1 and Y5. The coding regions of both genes are shown as black boxes, noncoding regions are hatched, and intron sequences are white. The exact locations and direction of transcription of the exons is indicated by arrows in the upper strand for the Y1 gene and in the lower strand for the Y5 gene. The alternatively spliced forms of both genes are shown above and below the two strands. **(B)** Schematic representation of exon 2 of the human Y5 gene. White boxes, putative transmembrane domain coding regions; hatched box, position of exon 1C of the Y1 gene.

NPY receptors also belong to this large superfamily of G-protein-coupled receptors; various cloning techniques have so far identified five receptor subtypes (Y1, Y2, Y4, Y5, and Y6) *(8)*. Sequence analyses of the cloned receptors reveal a substantial divergence in primary sequence between the receptor subtypes. The Y1 subtype was the first Y receptor to be cloned, and full-length cDNAs have been isolated from several species including human *(9)*, rat *(10)*, mouse *(11)*, and frog *(12)*. The Y1 receptors display a highly conserved structure, with overall identities of 94% or higher between mammalian species.

To determine the molecular organization and regulation of the human Y1 receptor gene, we isolated several genomic clones, subcloned and sequenced the exons and exon/intron boundaries, and mapped the transcription start sites (**Fig. 1**) *(13)*. In the course of this work, we identified several alternatively spliced 5' exons and also (surprisingly) an open reading frame in the opposite orientation within the most 5' exon, 1C *(14,15)*. The amino acid sequence showed significant homology to G-protein-coupled receptors, with the highest similarity (37%) to the Y1-receptor, suggesting that this new gene encodes a new Y-receptor subtype, the Y5 receptor. The Y5-receptor probably evolved by a gene duplication event from the Y1-receptor gene and is now encoded in the opposite orientation 23 kb upstream of exon 2 from the Y1 gene *(15)*. The transcription of both genes from opposite strands of the same DNA sequence suggests that transcriptional activation of one gene will have an effect on the regulation of expression of the other gene. As both Y1 and Y5 receptors are thought to play an important role in the regulation of food intake, coordinate expression of their specific genes may be important in the modulation of NPY activity *(16)*.

## 2. Materials

1. $[\alpha\text{-}^{32}P]$dCTP (Amersham #CA025687).
2. Agarose (Ultrapure; Life Technologies BRL #540-5510UA).
3. Cloning vectors: pBluescript (Stratagene) and pZero (Invitrogen).
4. Denhardt's solution (100X): 2% (wt/vol) Ficoll type 400 (Sigma [Castle Hill, NSW, Australia] #F2637); 2% (wt/vol) polyvinylpyrrolidene (Sigma PVP; P5288); 2% (wt/vol) bovine serum albumin (fraction V; Sigma #A7030) made up in RNase-free distilled $H_2O$. Filter through a Millipore 0.2-$\mu$m filter, and store in 10-mL aliquots at $-20°C$.
5. Ethylenediaminetetraacetic acid (EDTA; 0.5 *M*): make a 0.5 *M* solution (Sigma #E5134) in RNase-free distilled $H_2O$, stir and dissolve by adjusting to pH 8.0 with NaOH (10 *M*), and then autoclave.
6. Ethanol: absolute, analytical grade.
7. Ethidium bromide (EtBr) solution: make a 10-mg/mL solution (Sigma #E7637) in distilled $H_2O$; store at $4°C$ in a sealed, dark bottle protected against light (very hazardous/carcinogen).
8. $[\gamma\text{-}^{32}P]$ATP: NEN (Easytides™ #502211268).
9. Genomic library: Genome Systems, peripheral blood genomic P1 library (#FP1-2285).
10. Hexanucleotide mix: 0.06 $OD_{260}$ U/$\mu$L (Boehringer Mannheim #1277081) in 500 m*M* HEPES, pH 6.6, stored at $-20°C$.
11. Hybond N+: Amersham #RPN 203B.
12. Hybridization solution (1X): 6X standard saline citrate (SSC), 5X Denhardt's solution, 0.1% sodium dodecyl sulfate (SDS). Warm solution to $65°C$, and add denatured ssDNA to a final concentration of 100 $\mu$g/mL.
13. Klenow DNA polymerase: Boehringer Mannheim #1008404.
14. Kinase buffer (10X): 250 m*M* Tris-HCl, 50 m*M* $MgCl_2$, 25 m*M* 1,4-dithiothreitol (DTT) and 25% glycerol. Adjust pH to 9.5.
15. Lysis buffer: stock solution contains 50 m*M* D-glucose (Merck [Kilsyth, Victoria, Australia] #1-117.3Y), 10 m*M* EDTA (Sigma E5134) and 25 m*M* Tris-HCl. Adjust pH to 8.0, and store at $4°C$. Important: Add 50 mg/mL lysozyme (Boehringer Mannheim #1243004) just prior to use.

16. Phenol/chloroform (1:1) saturated with Tris borate-EDTA (TBE) buffer.
17. Phosphate buffered saline (PBS; 0.1 $M$): 10 m$M$ Na$_2$HPO$_4$ (Sigma S3264), 3 m$M$ KCl (Sigma P9541), 1.8 m$M$ KH$_2$PO$_4$ (Sigma P9791), 140 m$M$ NaCl (Sigma S3014). Adjust pH to 7.4 and autoclave. Stable for 2–3 weeks at 4°C.
18. Reaction buffer (10X): 500 m$M$ Tris-HCl, 50 m$M$ MgCl$_2$, 500 $\mu M$ dATP, 500 $\mu M$ dGTP, 550 $\mu M$ dTTP, 50 $\mu M$ dCTP, 1% (vol/vol) mercaptoethanol. Adjust pH to 8.0, and store in 5-$\mu$L aliquots at –20°C.
19. Restriction enzymes: Boehringer Mannheim.
20. RNaseA solution: dissolve ribonuclease A (RNase A; bovine pancreas; Sigma #R5503; stock stored –20°C) in distilled H$_2$O at a concentration of 25 mg/mL. Store aliquots at –20°C. Boil aliquot for 1 min to remove DNase activity, and use at a final concentration of 25 $\mu$g RNase/mL.
21. Salmon sperm sonicated ssDNA: Sigma #D9156; stored as a 10-mg/mL solution at –20°C. Just before use, denature this solution at 100°C for 5 min, and then rapidly cool on ice.
22. SDS: 20% solution; Sigma #L4522.
23. Sephadex slurry: equilibrate Sephadex G25 powder (Sigma #G-25-80) in 1X TE buffer (pH 8.0) overnight, store slurry at 4°C.
24. SSC (20X): 3 $M$ sodium chloride (Sigma #S3014), 0.3 $M$ Tri-sodium citrate (Sigma #C8532), in distilled H$_2$O, adjust PH to 7.0, and autoclave. Stable for 2–3 wk at room temperature.
25. T4 polynucleotide kinase: (Boehringer Mannheim #174645).
26. Tris-buffered saline (TBS); (10X): 1.5 $M$ NaCl, 1 $M$ Tris-HCl. Adjust pH to 7.5 with HCl.
27. TBE (20X): 0.9 $M$ Tris-borate, 0.08 $M$ EDTA.
28. X-ray film: X-Omat AR film (Kodak CAT 1651512 ).

## 3. Methods

### 3.1. Probe Isolation and Labeling

Y-receptor-specific probes are normally isolated by polymerase chain reaction (PCR) using subtype-specific oligo nucleotides as primers and cDNA-carrying plasmids as templates (*see* **Note 1**).

### 3.1.1. Y1 Receptor Probe Labeling (see **Note 2**)

1. Place 25–50 ng of template DNA into a microcentrifuge tube and add 5 $\mu$L of hexanucleotide primer solution and the appropriate volume of water to give a total volume of 50 $\mu$L (*see* **Note 3**).
2. Denature the double-stranded DNA by heating for 5 min in a boiling water bath or 95–100°C heating block.
3. Spin the tube briefly in a microcentrifuge to combine the solution again.
4. Place tube on ice and add 10 $\mu$L of the reaction buffer, 5 $\mu$L of the appropriate radiolabeled $\alpha^{32}$P-dNTP and 2 $\mu$L of Klenow enzyme (5 U/mL). Mix gently by

pipetting the solution up and down, and briefly spin in a microcentrifuge (*see* **Note 4**).

5. Incubate tube at 37°C for 10–30 min (*see* **Note 5**).
6. Stop the reaction by heating the sample for 5 min at 95–100°C, and then place on ice.

### 3.1.2. Y1 Receptor Probe Purification via Spun Column Procedure

*See* **Note 6**.

1. Plug a sterile 1-mL syringe barrel with glass fiber, packing it in place with the syringe plunger.
2. Slowly pipet 1 mL of Sephadex slurry into the column, taking care to avoid air bubbles.
3. Place the column in a screw-capped microcentrifuge tube, and spin this setup in a 15-mL disposable centrifuge tube at >1100*g* for 5 min.
4. Repeat the pipetting and spinning (**steps 2** and **3**) until the column content stabilizes at 1 mL of Sephadex G25.
5. Add a known volume (e.g., 100 µL) of water to the top of the Sephadex bed, place the column in a fresh microcentrifuge tube, and repeat **step 3**. The volume recovered should be equal to that added to the top of the column. Then the column is ready for use.
6. Add 50 µL of water to the labeled sample, and load the combined 100 µL onto the column. Place the column in a fresh microcentrifuge tube, and spin for 5 min at >1100*g*.
7. Measure the cpm values of the sample in a β-scintillation counter. The average cpm range of a sample prepared like this is about $2–5 \times 10^6$ cpm.
8. Store sample at –20°C (*see* **Note 7**).

## 3.2. Library Screening (17)

### 3.2.1. Checking for Crossreactivity of the Y1 Receptor Probe

To screen a high density P1 or BAC library successfully, it is important to check for crossreactivity of the probe first. Testing the probe against human genomic DNA, the P1 or BAC vector DNA, and a small test grid is recommended. Standard Southern blot analysis of restriction-digested human genomic DNA and restriction-digested P1 vector is also a good method. Restriction enzymes with a six-base recognition site such as *Eco*RI, *Bam*HI, or *Hind*III work well. If a high crossreactivity of the probe with the vector DNA or a high background in the genomic blot is observed, a smaller probe or a probe from a different portion of the gene should be considered. Alternatively, one can try to block the repeat elements with genomic and/or Cot-1 DNA competition. A slight background, barely revealing the individual spots on the test grid, is acceptable and can actually assist in the positioning of positive signals.

### 3.2.2. Prehybridization

1. For hybridizing eight filters (20 × 20 cm), make up 225 mL of hybridization solution consisting of:
   6X SSC
   5X Denhardt's solution
   0.5% SDS
   100 mg/mL ssDNA.
   Denature this solution just before use by heating it at 100°C for 5 min, and then rapidly cool on ice.
2. Heat the solution to 65°C in a suitable container able to fit the filters properly. A shaking water bath or an incubation oven with a shaking platform is used for this purpose.
3. Add the filters singly, avoiding trapping air bubbles between them, and incubate for 2 h at 65°C (*see* **Note 8**).

### 3.2.3. Hybridization

1. Take the filters out of the hybridization solution, and add $^{32}$P-labeled probe that has just been denatured by heating to 5 min at 100°C (*see* **Note 9**). For eight filters, a total of 400,000–600,000 cpm/mL is recommended, which normally equals two standard labeling reactions.
2. Add the filters back into the hybridization solution, again avoiding the trapping of air bubbles, and incubate overnight.
3. To remove nonspecifically bound probe, the following washes are carried out:
   300 mL of 2X SSC, 0.5% SDS at room temperature for 5 min.
   300 mL of 2X SSC, 0.1% SDS at room temperature for 15 min.
   300 mL of 0.1X SSC, 0.5% SDS at 37°C for 30 min.
   300 mL of 0.1X SSC, 0.5% SDS at 65°C for 30 min.
4. Place filters on Whatman filter paper to dry. Do not allow filters to dry completely. Otherwise, it will be difficult to carry out further washes (if necessary).
5. For autoradiographic detection, place filters between two pieces of plastic wrap in a cassette with two intensifying screens, and place X-ray film on top. Expose overnight at –80°C.
6. Store filters at –20°C (*see* **Notes 10** and **11**).

## 3.3. Clone Identification

As mentioned earlier, positive clones are clearly identifiable by the appearance of two discrete signals, oriented in a special way, within one square. A transparent replica of the grid is normally supplied by the manufacturer of the filters to assist in identification of the clone position on the autoradiogram (*see* **Note 12**). Two clones are normally enough for further analysis.

### 3.3.1. P1 Plasmid DNA Isolation

P1 clones are provided by the manufacturer as a bacterial stab culture. To isolate single clones, the bacteria must be streaked onto an agar plate containing 25 µg/mL kanamycin for selection and grown overnight at 37°C. P1 plasmids are normally maintained at one copy per cell; therefore a larger culture volume is required for the isolation of sufficient amounts of plasmid DNA.

1. Inoculate 30 mL of LB medium containing 25 µg/mL kanamycin with a single colony of bacteria, and incubate overnight at 37°C in a shaking incubator.
2. Collect the bacteria by centrifuging the culture at 10,000*g* for 5 min. Discard the supernatant.
3. Resuspend the bacteria in 1 mL of lysis buffer, transfer suspension to a 15-mL Falcon polypropylene tube, and incubate for 5 min at room temperature.
4. Add 2 mL of freshly prepared 0.2 *N* NaOH/1% SDS, seal the tube, invert the tube several times until the mixture becomes clear (do not vortex), and then place on ice for 5 min.
5. Add 1.5 mL of 3 *M* KAc pH 5.7, mix gently, and place on ice for 5 min.
6. Centrifuge for 5 min at 10,000*g*.
7. Transfer the supernatant to a new 15-mL Falcon tube, and add an equal amount of phenol/chloroform (1:1). Mix by repeated inversion, and centrifuge at 10,000*g* for 5 min.
8. Repeat step 7 twice.
9. Remove the aqueous phase, avoiding any contamination with the phenol/ chloroform, and place into a new falcon tube. Add precooled (–20°C) absolute ethanol to a final concentration of 70%, mix by inversion and centrifuge at 10,000*g* for 20 min at 4°C.
10. Remove the supernatant carefully, and wash the pellet with 3 mL of 70% ethanol. Centrifuge for 5 min at 10,000*g*.
11. Carefully discard the supernatant, and vacuum-dry the pellet until completely dry.
12. Resuspend the pellet in 100 µL of water containing 50 mg/mL of DNase-free RNase A. Transfer the DNA into a 1.5-mL microcentrifuge tube, and store at –20°C until further use.

## 3.4. Subcloning and Mapping

Direct sequencing of P1 plasmid DNA only works well using cycle sequencing. For exon/intron border determination and mapping, subcloning smaller fragments into a more suitable vector (e.g., pBluescript, pZero) is therefore recommended. This can easily be achieved by standard "shotgun" cloning of restriction enzyme-digested P1 plasmid DNA into the appropriate site of pBluescript or pZero. Six-base pair recognizing restriction enzymes such as *Eco*RI, *Hind*III, or *Bam*HI can be used to generate a small library of overlapping fragments. Cloning into the ampicillin-resistant pBluescript or the zeocin-resistant pZero vector avoids the survival of re-ligated kanamycin-resistant P1 plasmids. Individual white colonies can be picked and plasmid DNA isolated from them as described in **Subheading 3.3.1.** Alternatively, bacteria can be streaked out and screened with standard colony screening protocols, using specific probes (e.g., oligonucleotides) for a particular part of the cDNA, to obtain only those subclones that contain the portion of interest. Individual subclones are sequenced using standard techniques, and the sequences are compared with the cDNAs to determine the exon/intron borders.

## 3.5. Restriction Mapping With Oligonucleotide Probes

In this method the restriction map of large fragments subcloned from the P1 clone is determined by first digesting the plasmid with a rare cutting enzyme such as *Not*I or *Sfi*I to linearize the DNA, followed by partial digestion of the subsequent fragment with a more frequent cutting enzyme. The series of digestion products are then separated on an agarose gel, transferred to nitrocellulose, and probed with an oligonucleotide probe corresponding to one of the ends of the insert. The estimated difference in size of the hybridizing bands then corresponds to the size of the joining fragments. Maps for several different restriction enzymes can be obtained by this method; then they can be combined and verified by sequencing of the various identified fragments.

### 3.5.1. Transfer of DNA onto Hybond+

1. First soak the gel in 0.15 M HCl for 10 min to fragment the DNA. This aids in efficient transfer of DNA onto nitrocellulose. Agitate gently to ensure even exposure of gel to HCl. Alternatively, expose to ultraviolet (UV) light for 1 min. Caution must be used with the UV light approach, since the efficiency of nicking can decrease over time due to solarization of the filter. Efficiency of transfer should always be monitored by staining the gel following transfer.
2. If HCl is used, rinse gel with distilled $H_2O$ to remove excess acid.
3. Denature DNA by soaking gel in 1.5 $M$ NaCl and 0.5 $M$ NaOH for 30 min.
4. Neutralize by soaking for 30 min in 1 $M$ Tris-HCl, pH 8.0, containing 1.5 $M$ NaCl.
5. Place gel on 3–4 sheets of Whatman filter paper slightly larger than the gel. Saturate with 6X SSC. Briefly wet nitrocellulose with water, cut to the size of the gel, and then soak in 6X SSC. Place nitrocellulose and 1–2 sheets of saturated filter paper carefully on top of the gel followed by a stack of paper towels and a light weight to keep all layers compressed. Surround the gel with parafilm to prevent siphoning of transfer buffer onto paper towels.
6. Transfer should be complete in 8 h. The efficiency can be checked by restaining the gel after transfer in 100 mL $H_2O$, 1 mg/mL ethidium bromide.

### 3.5.2. Labeling of Oligonucleotide Probes

Label oligos with fresh $\gamma$-$^{32}$P-ATP. High-specific-activity $\gamma$ yields the best results.

1. Combine reagents as follows and perform the polynucleotide kinase reaction for 30 min at 37°C.

    2 µL 10X kinase buffer
    2 µL 10 m$M$ ATP
    1 µL oligonucleotide (100 ng)
    1 µL polynucleotide kinase (10 U)
    14 µL H$_2$O

2. Incubate at 68°C for 15 min to destroy the enzyme.
3. Unincorporated nucleotides can be removed with a spin column as described in **Subheading 3.1.1.**

### 3.5.3. Hybridization with Oligonucleotides

In general, the same conditions for the hybridization with oligonucleotides can be used as for double-stranded probes, with the exception of a reduced temperature. The hybridization temperature depends strongly on the length and the GC content of the oligonucleotide. Several formulas for calculating the theoretical "melting" temperature of the double strand complex of oligos have been developed and can be used to estimate the optimal temperature. However, experimental condition might have to be adjusted on a case by case basis. More information on all standard molecular biology techniques can be found in Sambrook et al. *(18)*.

## 4. Notes

1. Alternatively, DNA fragments of any particular Y receptor can also be isolated by restriction enzyme digests of plasmid DNA carrying the cDNA fragment and isolating the fragment after separation on an agarose gel.
2. Particular caution has to be taken in handling, storing, and disposing of radioactive material. Always wear gloves and safety glasses.
3. The reaction volume can be scaled up or down if more or less than 25 ng of DNA is to be labeled.
4. Avoid vigorous mixing of the reaction mixture, as this can cause severe loss of enzyme activity.
5. DNA can be labeled to high specific activity within 10 min at 37°C but can be labeled for longer periods of up to 3 h at the same temperature if an even higher level of labeling is desired.
6. Although probe purification is not always necessary for membrane applications, the removal of unincorporated label is recommended to reduce the background in filter hybridization experiments. Also, the use of a spin column is a quick and efficient way to achieve this goal.
7. Labeled probes should not be stored for long periods (maximum 3 days at –20°C): because of the short half-life of [32]P, substantial probe degradation can occur.
8. Incubation for shorter periods is possible particularly when the probe showed no signs of crossreactivity. However, a minimum of 30 min should be allowed. New

filters should be thoroughly soaked in hybridization solution and prehybridized for at least 2 h.

9. Place the filters on plastic foil, and proceed quickly to avoid cooling of the solution and filters.
10. The filters can be reused for hybridization many times. We have successfully used one set 18 times. However, in order to achieve this, one has to reduce handling of the filters by not stripping the filter between hybridizations. Repetitive hybridizations with different probes will show new hybridizing clones at each step.
11. Filters should be wrapped in plastic wrap and stored at –20°C between hybridizations. Before reusing filters, thaw them completely at room temperature prior to removing the plastic wrap.
12. Although identification of the position of positive clones is straightforward, obtaining a second opinion from a colleague to confirm the exact grid position is recommended.

## References

1. Dilella, A. G. and Woo, S. L. C. (1987) Cloning large segments of genomic DNA using cosmid vectors. *Methods Enzymol.* **152**, 199–212.
2. Wahl, G. M., et al. (1987) Novel cosmid vectors for genomic walking, restriction mapping and gene transfer. *Proc. Natl. Acad. Sci. USA* **84**, 2160–2164.
3. Lozovskaya, E. R., Petrov, D. A., and Hartl, D. L. (1993) A combined molecular and cytogenetic approach to genome evolution in Drosophila using large-fragment DNA cloning. *Chromosoma* **102,** 253–266.
4. Wang, M., Chen, X. N., Shouse, S., Manson, J., Wu, Q., Li, R., et al. (1994) Construction and characterisation of a human chromosome 2-specific BAC library. *Genomics* **24**, 527–534.
5. Bockaert, J. (1991) G proteins and G-protein-coupled receptors: structure, function and interactions. *Curr. Opin. Neurobiol.* **1,** 32–42.
6. Maeda, N. and Smithies, O. (1986) The evolution of multigene families: human haptoglobin genes. *Annu. Rev. Genet.* **20,** 81–108.
7. Soares, M. B., Schon, E., Henderson, A., Karathanasis, S. K., Cate, R., Zeitlin, S., et al. (1985) RNA-mediated gene duplication: the rat preproinsulin I gene is a functional retroposon. *Mol. Cell. Biol.* **5,** 2090–2103.
8. Blomqvist, A. G. and Herzog, H. (1997) Y-receptor subtypes—how many more? *Trends Neurosci.* **20**, 294–298.
9. Herzog, H., Hort, Y. J., Ball, H. J., Hayes, G., Shine, J., and Selbie, L. A. (1992) Cloned human neuropeptide Y receptor couples to two different second messenger systems. *Proc. Natl. Acad. Sci. USA* **89,** 5794–5798.
10. Krause, J., Eva, C., Seeburg, P. H., and Sprengel, R. (1992) Neuropeptide Y1 subtype pharmacology of a recombinantly expressed neuropeptide receptor. *Mol. Pharmacol.* **41,** 817–821.
11. Eva, C., Oberto, A., Sprengel, R., and Genazzani E. (1992) The murine NPY-1 receptor gene. Structure and delineation of tissue-specific expression. *FEBS Lett.* **314,** 285–288.

12. Blomqvist, A. G., Roubos, E. W., Larhammar, D., and Martens, G. J. (1995) Cloning and sequence analysis of a neuropeptide Y/peptide YY receptor Y1 cDNA from *Xenopus laevis*. *Bioch. Biophys. Acta.* **1261,** 439–441.

13. Herzog, H., Baumgartner, M., Vivero, C., Selbie, L. A., Auer, B., and Shine, J. (1993) Genomic organisation, localisation, and allelic differences in the gene for the human neuropeptide Y Y1 receptor. *J. Biol. Chem.* **268,** 6703–6707.

14. Ball, H. J., Shine, J., and Herzog, H. (1995) Multiple promoters regulate tissue-specific expression of the human NPY-Y1 receptor gene. *J. Biol. Chem.* **270,** 27,272–27,276.

15. Herzog, H., Darby, K., Ball, H., Hort, Y., Beck-Sickinger, A., and Shine, J. (1997) Overlapping gene structure of the human neuropeptide Y receptor subtypes Y1 and Y5 suggests coordinate transcriptional regulation. *Genomics* **41,** 315–319.

16. Gerald, C., Walker, M. W., Criscione, L., Gustafson, E. L., Batzl-Hartmann, Ch., Smith, K. E., et al. (1996) A receptor subtype involved in neuropeptide Y induced food intake. *Nature* **382,** 168–171.

17. Shovlin, C. L. (1996) Streamlined procedures for screening a P1 library. *Biotechniques* **21,** 388–390.

18. Sambrook, J., Fritsch, E. F., and Maniatis, T. (1992) *Molecular Cloning (A Laboratory Manual),* 2nd ed. Cold Spring Harbor Laboratory Press, Cold Spring Harbor, NY).

# 3

# Human Type 2 Neuropeptide Y Receptor Gene

*Isolation and Characterization*

## David A. Ammar and Debra A. Thompson

## 1. Introduction

Our interest in G-protein signaling events involved in the specialized functions of the retinal pigment epithelium led us to investigate the pattern of G-protein-coupled receptor expression in this cell type. We used reverse transcriptase-coupled polymerase chain reaction (PCR) to amplify cDNAs derived from bovine tissue using degenerate oligonucleotide primers corresponding to conserved sequences in receptor transmembrane helices two and seven (*1*). One of the sequences identified encoded the partial sequence of a novel protein most closely related to the type 1 neuropeptide Y (NPY) receptor, the only member of the NPY receptor family characterized by molecular cloning at that time (*2,3*). To identify DNA sequences encoding the corresponding full-length protein necessary to verify NPY receptor activity, we chose to clone the human genomic DNA (*4*). This approach offers a number of advantages over cDNA library screening, especially for identification of clones corresponding to low-abundance transcripts. In a genomic DNA library, each gene is represented equally, whereas representation in a cDNA library is proportional to expression level. In addition, the large insert size favors the possibility that one clone will encompass the entire gene and allows the gene structure to be determined. Furthermore, the large genomic DNA fragments obtained are suitable for use in determining chromosomal localization by hybridization analysis of somatic cell hybrid panels and *in situ* hybridization of metaphase chromosomes.

From: *Methods in Molecular Biology, vol. 153: Neuropeptide Y Protocols*
Edited by: A. Balasubramaniam © Humana Press Inc., Totowa, NJ

## 2. Materials

Molecular biology grade reagents and Milli-Q water should be used for all procedures. Sources, where listed, are recommendations. Unless otherwise noted, solutions may be made ahead of time and stored at room temperature.

### 2.1. Plating and Screening the Genomic Library

1. Phage library: human genomic DNA library in λ DASH II (Stratagene).
2. Bacteria: *Escherichia coli* strain XL1-Blue MRA(P2) (Stratagene) is used as a host for initial screening. Strain LE392 is used in second- and third-round screenings and for phage lysates. This strain yields a high titer of phage but is not *recA–*.
3. 1 $M$ MgSO$_4$ (sterile filtered).
4. 20% (w/v) maltose (sterile filtered).
5. NZYM medium: dissolve 10 g NZ amine, 5 g yeast extract, 5 g NaCl, and 2 g MgSO$_4$·7H$_2$O in 1 L water, adjust pH to 7.4 with NaOH, and autoclave. Store tightly capped at 4°C.
6. NZYM agar plates: Add 15 g Bacto-agar (Difco) to 1 L of NZYM medium, and autoclave. Cool to 65°C in a water bath, and fill 10- and 15-cm Petri dishes about half-full. Store at 4°C (*see* **Note 1**).
7. NZYM top agarose: add 1.75 g electrophoresis-grade agarose to 250 mL NZYM medium, and autoclave. Store tightly capped at 4°C.
8. Luria-Bertani (LB) medium: dissolve 10 g Bacto-tryptone, 5 g yeast extract, and 10 g NaCl in 1 L water. Adjust pH to 7.4 with NaOH, and autoclave. Store tightly capped at 4°C.
9. 1 $M$ tris(hydroxymethyl)aminomethane-HCl (Tris-HCl), pH 7.5 and pH 8.0.
10. 0.5 $M$ ethylenediamine tetraacetic acid (EDTA), pH 8.0.
11. SM solution: 0.58 g NaCl, 0.2 g MgSO$_4$·7H$_2$O, 5 mL 1 $M$ Tris-HCl, pH 7.5, and 2 g gelatin in 100 mL. Autoclave, and store tightly capped at 4°C.
12. 20X sodium chloride/sodium citrate (SSC): 175.3 g NaCl and 88.2 g sodium citrate in 1 L. Adjust pH to 7.0 with NaOH.
13. Denaturing solution (0.5 $M$ NaOH, 1.5 $M$ NaCl): 20 g NaOH and 87.7 g NaCl in 1 L.
14. Neutralizing solution (0.5 $M$ Tris-HCl, pH 8.0, 1.5 $M$ NaCl): 60.59 g Trizma base and 87.7 g NaCl in 1 L. Adjust pH to 8.0 with HCl.
15. Rinse solution: 5X SSC, 0.5% sodium dodecyl sulfate (SDS).
16. 50X Denhardt's solution: 5 g Ficoll, 5 g polyvinylpyrrolidone, and 5 g bovine serum albumin (Fraction V) in 500 mL. Filter through Whatman filter paper on a Buchner funnel. Dispense into 50-mL aliquots and store at –20°C.
17. Prehybridization solution: 6X SSC, 1X Denhardt's solution, 0.1% SDS.
18. 10% (w/v) SDS.
19. Chloroform.
20. 10- and 15-cm sterile Petri dishes.
21. 15-mL polypropylene centrifuge tubes with snap-tops and screw-caps.
22. 50-mL polypropylene centrifuge tubes with screw-caps.

23. 85- and 137-mm round nitrocellulose transfer membranes (Millipore Immobilon-NC membranes, catalog nos. HATF 085 50 and HATF 137 50).
24. Whatman 3MM chromatography paper.
25. $18^1/_2$-gage needles and ink-filled syringe.
26. Ultraviolet (UV) crosslinker (Stratagene).
27. Small and large plastic containers for hybridization and washing of library lifts.
28. 1.5-mL microcentrifuge tubes, autoclaved.
29. $5^1/_2$-inch Pasteur pipets, autoclaved.

## 2.2. Preparation of Radiolabeled cDNA Probe

1. Agarose and low-melting-point agarose.
2. 50X Tris-sodium acetate (TSA): dissolve 242 g Trizma base, 34 g sodium acetate·$3H_2O$, and 80 mL of glacial acetic acid in water. Adjust final volume to 1 L and pH to 8.0.
3. 50X Tris-acetate-EDTA (TAE): dissolve 242 g Trizma base, 57 mL glacial acetic acid, and 100 mL of 0.5 $M$ EDTA in water. Adjust final volume to 1 L and pH to 8.0.
4. 5 mg/mL ethidium bromide. Store in dark bottle.
5. 3 $M$ sodium acetate, pH 5.2.
6. 0.75 $M$ sodium phosphate, pH 3.5.
7. Sephadex G-50 (50–150 µm) in 1X SSC, autoclaved.
8. Ethanol.
9. Autoclaved water.
10. Rediprime™ random oligonucleotide labeling kit (Amersham).
11. Redivue [$\alpha$-$^{32}$P]dCTP: 3000 Ci/mmol, 10 mCi/mL (Amersham).
12. PEI-cellulose thin-layer chromatography plates.
13. Electrophoresis chamber and power supply.
14. Ultra-free MC spin columns (Schleicher and Schuell).
15. 1-mL tuberculin syringes with porous Teflon frits or glass wool.
16. Kodak X-Omat film and cassettes.

## 2.3. Subcloning Genomic Insert DNA

1. Solutions and materials from above Subheadings.
2. 10 mg/mL RNase A.
3. 10 mg/mL DNase 1.
4. 25 mg/mL Proteinase K.
5. 5 $M$ LiCl, autoclaved.
6. Phenol equilibrated with 10 m$M$ Tris-HCl, pH 8.0. Store at 4°C in dark bottle.
7. Chloroform/isoamyl alcohol (24:1).
8. Phenol:chloroform:isoamyl alcohol (25:24:1). Store at 4°C in dark bottle.
9. TE: 10 m$M$ Tris-HCl, pH 7.5, 1 m$M$ EDTA.
10. *Not*I restriction endonuclease and reaction buffer (New England Biolabs).
11. Plasmid pGEM-5Zf(-) (Promega).

12. T4 DNA ligase and reaction buffer (New England Biolabs).
13. JM109 competent cells and supplied SOC medium (Promega).
14. LB/Amp/X-gal/IPTG agar plates: add 15 g of Difco agar to 1 L of LB and autoclave. Cool to 60°C in a water bath, and add ampicillin to 50 mg/L final concentration. Pour into 10 cm Petri dishes (*see* **Note 1**). Before use, coat surfaces with 100 µL of sterile-filtered 40 m*M* isopropyl β-D-thiogalactopyranoside (IPTG) in water and 100 µL of 2 mg/mL 5-bromo-4-chloro-3-indolyl β-D-galactopyranoside (X-gal) in dimethyl formamide.
15. Solution I: 50 m*M* glucose, 25 m*M* Tris-HCl, pH 8.0, 10 m*M* EDTA. Prepare fresh.
16. Solution II: 0.2 *M* NaOH, 1% SDS. Make fresh weekly.
17. Solution III: 60 mL 5 *M* potassium acetate, 11.5 mL glacial acetic acid, 28.5 mL water.
18. Lysozyme.
19. CsCl, ultracentrifugation grade or better.
20. Cheesecloth.
21. *n*-butanol saturated with TE and CsCl.
22. Dialysis buffer: 100 m*M* Tris-HCl, pH 7.5, 10 m*M* EDTA, 0.5 *M* NaCl.
23. Dialysis tubing (SpectraPor, molecular weight cutoff 12-14 kDa).
24. Corex glass centrifuge tubes, 15 and 30 mL.
25. SW41 rotor and polycarbonate centrifuge tubes (Beckman).
26. TLA-100.3 rotor and Quick-seal centrifuge tubes (Beckman).

## *2.4. Mapping Insert DNA; Identifying 5'-Untranslated Region*

1. Solutions and equipment from above sections.
2. Restriction endonucleases (six-cutters, e.g., *Apa*I, *Bam*HI, *Eco*RI, *Hind*III, *Bgl*II, *Pst*I, *Sca*I, and *Pvu*II).
3. Oligonucleotide primers corresponding to sequences upstream of the translation initiation codon.
4. cDNA probes corresponding to various regions of the transcript.
5. Diethyl pyrocarbonate (DEPC).
6. DEPC-treated water. Add 1 mL DEPC to 1 L Milli-Q water, incubate overnight, and then autoclave twice for 1 h to remove excess DEPC.
7. Human hippocampal poly(A) RNA (Clonetech).
8. dATP, dCTP, dGTP, and dTTP: 100 m*M* stock solutions (Gibco/BRL).
9. RNasin (Promega).
10. Moloney murine leukemia virus (M-MLV) reverse transcriptase and supplied reaction buffer (Gibco/BRL).
11. Taq DNA polymerase and supplied reaction buffer (Gibco/BRL).
12. [γ-$^{32}$P]ATP: 10 mCi/mL, 3000 Ci/mmol (Amersham).
13. Polynucleotide kinase (PNK) and supplied reaction buffer (New England Biolabs).
14. 20X DET: 2 g polyvinylpyrrolidone, 2 g Ficoll 400, 2 g bovine serum albumin (BSA); (Fraction V), 2 mL 1 *M* Tris-HCl, pH 8.0, and 4 mL 0.5 *M* EDTA in 500 mL.
15. Oligo prehybridization solution: 5 mL 20X SSC, 25 mL 20X DET, 2.5 mL 10% SDS, and 17.5 mL water.

16. Thin-walled PCR tubes.
17. Capillary transfer system.
18. Hybond-N (Amersham).
19. Sealable plastic bags and heat sealer.

## 2.5. Expression and Ligand Binding Studies of Recombinant Protein

1. HEK293 cells (ATTC, CRL 1573).
2. Dulbecco's modified Eagle's medium (DMEM) containing penicillin/streptomy-cin and 10% fetal bovine serum. Store at 4°C.
3. Trypsin-EDTA: 0.5 g trypsin and 0.02 g $Na_4EDTA$ per liter.
4. DEAE-dextran: 40 mg/mL in DMEM. Sterile filter and store at 4°C.
5. Chloroquine, 100 m$M$ . Sterile filter and store at 4°C.
6. Phosphate-buffered saline (PBS) containing 10% dimethylsulfoxide (DMSO). Sterile filter.
7. Protease inhibitors (final concentrations): 5 m$M$ benzamidine, 0.1 m$M$ phenyl-methylsulfonyl fluoride (PMSF), and 2 µg/mL each of aprotinin, pepstatin, and leupeptin.
8. Working buffer: 66.7 m$M$ sodium phosphate, pH 7.0.
9. Hypotonic buffer: 1/10 dilution of working buffer.
10. Sucrose: 15 and 40% (w/v) in working buffer containing protease inhibitors.
11. Membrane buffer: 50 m$M$ HEPES, pH 7.4, 5 m$M$ $MgCl_2$, 1 m$M$ $CaCl_2$.
12. NPY, peptide YY (PYY), pancreatic polypeptide (PP), $NPY_{13-36}$, [$Leu^{31}$, $Pro^{34}$]NPY, substance K, substance P, and neuromedin K (Sigma or Peninsula Laboratories).
13. Bolton-Hunter labeled [$^{125}$I]NPY: 2200 Ci/mmol (New England Nuclear/Dupont).
14. Binding buffer: Membrane buffer containing 0.5% BSA and 0.05% bacitracin.
15. Dye reagent for protein assays (Bio-Rad).
16. 15-cm tissue culture dishes.
17. Teflon-glass homogenizer.
18. 50 mL polypropylene centrifuge tubes with caps.
19. $12 \times 75$ mm polypropylene centrifuge tubes.
20. SW28 rotor and polyallomer centrifuge tubes (Beckman).
11. Ti70 rotor and polycarbonate centrifuge tubes (Beckman).
22. Vacuum filtration manifold (Millipore).

# 3. Methods

## 3.1. Plating and Lifting the Genomic DNA Phage Library

For all procedures, use sterile techniques and autoclaved solutions. Follow standard biohazard protocols when working with recombinant DNA and bacteria.

1. The day prior to plating, inoculate 5 mL of LB supplemented with 0.2% maltose and 10 m$M$ $MgSO_4$ with host bacteria (XL1-Blue MRA [P2]) from a glycerol stock. Grow the cells overnight in a 37°C shaking incubator.

2. The following morning, add 0.2 mL of the overnight culture to 20 mL fresh LB containing 0.2% maltose and 10 m$M$ MgSO$_4$, and grow the cells until the OD$_{600nm}$ reaches 0.5.

3. Collect the cells by centrifugation in a clinical centrifuge at RT.

4. Resuspend the cells in 20 mL sterile 10 m$M$ MgSO$_4$ on ice.

5. Warm NZYM agar plates in a 37°C incubator.

6. Liquify NZYM top agarose in a microwave, and then place in a 45°C water bath.

7. For each 15-cm NZYM agar plate, dispense 0.6 mL of bacteria into a sterile 15-mL centrifuge tube on ice.

8. Dilute an aliquot of the genomic DNA phage library in SM to a final concentration of 5000 pfu/µL and place on ice.

9. Add 5 µL of diluted phage to each tube containing bacteria, and gently mix. Prepare the tubes in batches of two to three, staggered by a few min.

10. Prepare negative control (5 µL of SM buffer) and positive control mixtures (*see* **Note 2**).

11. Transfer tubes in batches to a 37°C water bath, and incubate for 15 min.

12. Add 7.5 mL of NZYM top agarose to the first tube, and pour the contents onto a warm NZYM plate.

13. Shake the plate to allow the agarose to cover the surface completely. Any air bubbles that form can be popped with a glass pipet tip heated in a Bunsen burner.

14. Repeat for the rest of the samples.

15. After the plates have hardened, incubate agar side up at 37°C for 8–12 h, until plaques are visible (about 1 mm in diameter with clear centers).

16. Chill the plates at 4°C.

17. Place a sheet of Whatman 3MM filter paper into a large tray, saturate it with denaturing solution, and pour off the excess liquid (Tray 1). Prepare a second tray containing Whatman 3MM paper saturated with neutralizing solution (Tray 2).

18. Place a large dish containing 300 mL of rinse solution on an orbital shaker at slow speed.

19. Wearing gloves, label a 137-mm nitrocellulose membrane with a permanent marker.

20. Using forceps, place the membrane on the surface of a cold NZYM plate, being careful not to trap air pockets between the membrane and the agarose.

21. Mark the orientation of the membrane on the plate by piercing at several places along the edge with an 18$^1$/$_2$-gage needle and ink-filled syringe.

22. After 2 min, carefully remove the membrane from the plate using forceps, and place it plaque side up in Tray 1 for 4 min.

23. Transfer the membrane to Tray 2 for 5 min.

24. Transfer the membrane to the rinse solution. Membranes can be left at this step until lifting is complete.

25. Duplicate lifts and lifts of the remaining plates are generated by repeating **steps 19–24** (*see* **Note 3**).

26. Place the membranes on Whatman 3MM paper plaque side up, and UV crosslink according to the manufacturer's instructions.

## 3.2. Preparation of Radiolabeled Probe

The cDNA insert can be purified in advance, while the actual labeling reaction should be performed on the same day as the hybridization. Prepare one Rediprime labeling reaction for every six membrane lifts to be hybridized.

1. Release the insert from cloned type 2 NPY receptor cDNA (e.g., a clone derived from an initial RT-PCR product) by restriction digestion. Electrophorese the digest on a low-melt agarose gel in 1X TSA.
2. Stain the gel for 10 min in 1X TSA containing 0.5 µg/mL ethidium bromide. Cut out the insert cDNA band using a razor blade and UV-transilluminator.
3. Transfer the band to a sterile microcentrifuge tube, freeze and thaw it, and then break it into small pieces with a heat-sealed glass pipet tip.
4. Transfer the pieces to the top chamber of an Ultra-free MC centrifuge filtration device and spin at 10,000$g$ for 30 min.
5. Add 100 µL of TE to the top chamber, and spin again.
6. To the combined filtrate, add 1/10 volume of 3 $M$ sodium acetate, pH 5.2, and 3 volumes of ethanol, and incubate at –20°C for 1 h to precipitate DNA.
7. Pellet the DNA by centrifuging at 10,000$g$ for 30 min at 4°C. Rinse the pellet with ice-cold 70% ethanol and air dry.
8. Resuspend the pellet in 50 µL water, and estimate the cDNA concentration by running an aliquot on a 1X TAE-agarose gel containing 0.1 µg/mL ethidium bromide (*see* **Note 4**).
9. Add 25 ng insert cDNA in 45 µL water to each Rediprime tube, and resuspend the lyophilized pellet.
10. Add 5 µL [$\alpha$-$^{32}$P]dCTP, mix, and incubate at 37°C for 30 min, and then at 68°C for 10 min.
11. Spot 0.5 µL of the reaction mix on a small rectangle of PEI-cellulose about 1 cm from the lower edge and air dry.
12. Place the PEI-cellulose in a beaker containing 0.5 cm of 0.75 $M$ sodium phosphate, pH 3.5, with the spot just above the level of the liquid.
13. Allow the buffer to migrate about 5 cm up the PEI-cellulose, then remove from the beaker, and air dry.
14. Wrap the PEI-cellulose in plastic wrap, expose to film for 5 min, and develop. Probe with [$\alpha$-$^{32}$P]dCTP incorporated will remain at the origin, while free radioactivity will migrate near the buffer front. Use probes with >70% incorporation.
15. Plug the end of a 1-mL tuberculin syringe with glass wool or a porous Teflon frit, and fill the syringe to the 1-mL mark with Sephadex-G50 in 1X SSC.
16. Allow excess 1X SSC to drain through, and then suspend the syringe over a microcentrifuge tube.
17. Dilute labeling reactions to 300 mL with 1X SSC, and load each reaction on a Sephadex-G50 column. Collect the eluates (fraction #1).
18. Add 300 mL of 1X SSC to the top of the columns, and collect the eluates in fresh tubes (fraction #2).
19. Repeat the elution two more times, collecting fractions #3 and #4.

20. Analyze all four fractions by PEI-cellulose chromatography as described above. Most of the labeled probe should be in fraction #2.

### *3.3. Hybridization and Washing Library Lifts*

### *3.3.1. Screening Primary Lifts*

1. Warm the prehybridization solution to 65°C.
2. Place the membrane library lifts in round plastic containers containing 300 mL of prehybridization buffer. Incubate for 3 h at 65°C in a shaking water bath.
3. Heat the radiolabeled probe from **Subheading 3.2.** for 5 min at 95°C in a microcentrifuge tube with screw-cap lid. Quick-chill on ice.
4. Add the probe to fresh prehybridization solution, preparing 2 mL for each membrane.
5. Place the membranes, one by one, into containers containing the probe solution (no more than 10 membranes each), and hybridize overnight at 65°C in a shaking water bath.
6. Remove the hybridization solution, and store at –20°C for use in subsequent screenings.
7. Wash the membranes in two changes of 2X SSC, 0.1% SDS at 65°C for 10 min using 300 mL per 10 membranes.
8. Perform up to three additional washes in 1X SSC and 0.1% SDS at 65°C. Monitor both positive and negative control lifts with a pancake-type detector. Stop washing when the radioactivity on the negative lift is near background and the radioactivity on the positive lift is greater than that on the negative lift.
9. Wrap the membranes in plastic wrap and tape to filter paper. Place fluorescent or radioactive orientation markers at various places on the filter paper.
10. Expose to X-ray film for 2–3 days.
11. Orient films over the membranes by aligning the orientation markers and needle marks, and outline the membranes with a marker pen.
12. Place the films of the duplicate lifts on top of one another, and locate hybridization signals that appear in the same position on both lifts. Circle with a marker pen.
13. Align the films with the NZYM plates using the ink spots. Circle the positions of positive plaques on the back of the plates.
14. Core the positive plaques using the large end of a sterile glass pipet.
15. Transfer the plugs to sterile centrifuge tubes containing 0.5 mL of SM and a drop of chloroform, vortex, and incubate at RT for 1 h or overnight at 4°C.

### *3.3.2. Second and Third Round Screening*

1. Prepare the bacterial host LE392, top agarose, and 10 cm NZYM plates as described in **Subheading 3.1.**
2. Estimate the titer of the SM containing the primary phage plugs and prepare 3 dilutions at roughly 10, 50, and 100 pfu/μL (*see* **Note 5**).
3. For each 10-cm NZYM agar plate, dispense 0.2 mL of bacteria into a sterile 15-mL centrifuge tube on ice.

4. Add 5 µL of each dilution to the aliquots of bacteria and proceed with **steps 11–16** (*see* **Subheading 3.1.**) using 4 mL of warmed top agarose per tube.
5. Prepare lifts from plates that have 100–500 plaques (the density of the plaques should be low enough so that they do not touch one another) as described in **Subheading 3.1., steps 17–26**, using 87-mm membranes.
6. Prehybridize the membranes as described above, and then hybridize with the probe solution from the primary screening that has been heated for 10 min at 95°C and cooled to 65°C.
7. Wash the membranes as described in **Subheading 3.3.1.**, with final washes at 0.1X SSC and 0.1% SDS.
8. Expose the lifts to film, and identify hybridization signals and positive plaques as described above. Core the middle of each plaque using the small end of a glass pipet, and transfer to a centrifuge tube containing 0.1 mL SM and a drop of chloroform.
9. Estimate the titer of the secondary plugs, and prepare two or three dilutions ranging between 5 and 50 pfu/µL.
10. Prepare NZYM plates containing 25–50 plaques, and repeat the lifting and screening as described above. All third-round plaques should appear positive in the hybridization. If not, perform a fourth-round screen.

## 3.4. Subcloning Genomic DNA Inserts and Plasmid DNA Preparation

### 3.4.1. Preparation of Phage DNA from Liquid Lysates

1. Prepare bacterial host LE392 as in **Subheading 3.1.** and place 0.2-mL aliquots in 50-mL screw-cap centrifuge tubes on ice.
2. Estimate the titer of phage plugs in SM that were positive in the third-round screen (*see* **Note 5**). For each, add a range of phage (between $10^4$ and $10^5$ pfu) to each of three tubes of bacterial host.
3. Incubate the tubes for 15 min in a 37°C water bath, and then add 10 mL NZYM medium to each tube.
4. Place caps loosely on the tubes, and incubate in a 37°C shaking incubator for 8–12 h. The cultures will become turbid over the first few hours, and then will gradually clear. Stop the incubation when the cultures have cleared and before regrowth of bacterial host.
5. Choose the tubes showing the best lysis, add a few drops of chloroform, shake, and let stand for 5 min or store overnight at 4°C.
6. Centrifuge the lysates at 1250*g* for 10 min, and transfer the supernates to fresh tubes. Reserve 5 mL from each as a phage stock, and store at 4°C.
7. To the remaining lysates, add 2 µL DNase I and 2 µL RNase A, and incubate for 1 h at 37°C.
8. Add 200 µL Proteinase K solution, and incubate for 15 min at 37°C.
9. Add 0.25 mL 0.5 *M* EDTA and 0.25 mL 10% SDS, and incubate at 65°C for 1 h.
10. Centrifuge at 1250*g* for 10 min, and transfer the supernatants into SW41 centrifuge tubes.

11. Add 1 mL 5 *M* LiCl and 6 mL isopropanol to each tube, and incubate at –20°C for 1 h.
12. Centrifuge at 70,000*g* in an SW41 rotor for 20 min at 4°C.
13. Resuspend the pellets in 0.4 mL TE, and transfer to microcentrifuge tubes.
14. Extract twice with phenol:chloroform:isoamylalcohol, and then twice with chloroform/isoamylalcohol.
15. Ethanol-precipitate the DNA, and wash the pellet as described in **Subheading 3.2.**
16. Resuspend the pellet in 50 μL TE, and determine the DNA concentration by the 260-nm absorbance (expected yield is 10–40 μg).

## 3.4.2. Ligations, Transformations, and Plasmid DNA Minipreps

1. Digest phage DNA by incubating an aliquot with *Not*I (5 U enzyme/μg DNA) overnight at 37°C. Determine insert sizes by running aliquots of the digests on an agarose gel in 1X TAE containing 0.1 μg/mL ethidium bromide.
2. Digest pGEM-5Zf(-) with *Not*I, and check the reaction on an agarose gel.
3. Add together 100 ng linearized pGEM-5Zf(-) and 4.5 μg *Not*I-digested phage DNA in a microcentrifuge tube (*see* **Note 6**). Adjust the volume to 100 μL with water, and precipitate the DNA as described in **Subheading 3.2.**
4. Resuspend the DNA in 17 μL autoclaved water, add 2 μL 10X T4 ligase buffer and 1 μL T4 ligase, and incubate overnight at 15°C.
5. Thaw JM109 competent cells on ice and place 0.1-mL aliquots in sterile 15-mL snap-top tubes on ice.
6. Add 1 mL ligation reaction to the cells, and incubate on ice for 10 min.
7. Heat-shock the cells for 45 s in a 42°C water bath, and then incubate on ice for 2 min.
8. Add 0.9 mL warm SOC to each tube, and place in a 37°C shaking incubator for 1 h.
9. Plate various amounts of the ligation mix on LB/Amp/X-gal/IPTG plates, and incubate overnight at 37°C. White colonies grow from bacteria containing plasmid with insert.
10. Restreak several white colonies on fresh LB/Amp/X-gal/IPTG plates.
11. Inoculate 5-mL aliquots of LB/Amp with single, white colonies using a sterile loop. Grow overnight in a 37°C shaking incubator.
12. Transfer 1.5 mL of the cultures to microcentrifuge tubes, and centrifuge to pellet the cells. Store the remaining cultures at 4°C.
13. To the cell pellets, add 200 μL of solution I containing 2 mg/mL lysozyme, and resuspend by vortexing.
14. Add 200 μL of solution II, and incubate on ice for 5 min.
15. Add 200 mL of 3 *M* sodium acetate, pH 5.2, mix, and centrifuge at 10,000*g* for 1 min.
16. Transfer the supernatants to new microcentrifuge tubes, and ethanol-precipitate the DNA as described in **Subheading 3.2.**
17. Resuspend the pellets in 50 μL TE containing 0.1 μg/mL RNase A.
18. Check for inserts by digesting 5 μL DNA with restriction enzymes flanking the *Not*I site in the multicloning site of pGEM5Zf(-) and running the digests on agarose gels as described above.

### 3.4.3. Large-Scale Preparation of Plasmid DNA

1. Inoculate 200 mL TB/Amp with cultures containing genomic DNA inserts, and incubate overnight at 37°C with shaking.
2. Pellet the bacteria by centrifuging at 4000*g* for 20 min at 4°C.
3. To each cell pellet, add 10 mL of solution I containing 40 mg of lysozyme, and resuspend by vortexing. Incubate at RT for 10 min with shaking.
4. Add 20 mL of solution II, and incubate for 10 min on ice.
5. Add 10 mL of ice-cold solution III, and incubate for 10 min on ice with shaking.
6. Centrifuge at 10,000*g* for 25 min at 4°C, and strain the supernate through 4 layers of cheese cloth into 30 mL Corex centrifuge tubes.
7. Add 0.6 volumes of isopropanol (about 20 mL), and incubate on ice for 5 min.
8. Centrifuge at 10,000*g* for 25 min at 4°C, discard the supernatant, and invert the tubes on paper towels to drain.
9. Resuspend the DNA pellets in 5 mL of TE by vortexing. Warm the solution to 37°C if necessary.
10. Add 10 μL of RNase A, and incubate at 37°C for 15 min. Transfer to 15-mL screw-top tubes.
11. Extract once with an equal volume of phenol:chloroform:isoamylalcohol and twice with an equal volume of chloroform:isoamylalcohol.
12. Transfer DNA solutions to 15-mL Corex tubes, add 0.5 mL 3 *M* sodium acetate, pH 5.2, and 10 mL ethanol, and incubate on ice for 30 min.
13. Centrifuge at 10,000*g* for 25 min at 4°C, discard the supernatant, and drain the tubes.
14. Resuspend pellets in 2.6 mL TE, and then add 0.1 mL ethidium bromide and 2.6 g CsCl.
15. Transfer the DNA solutions to 3.5-mL Quick-seal tubes using $18^1/_2$-gage needles and syringes. Fill tubes halfway up the neck, place metal caps on the tubes, and heat-seal.
16. Centrifuge at 240,000*g* in a TLA-100.3 rotor for 18 h at 20°C.
17. Insert $18^1/_2$ gage needles into the tops of the tubes, and collect the lowest heavy red band in each gradient by piercing through the side of the tubes using $21^1/_2$-gage needles with a syringe.
18. Transfer the DNA to 1.5-mL microcentrifuge tubes, add an equal volume of CsCl/TE-saturated butanol, vortex, and centrifuge at 12,000*g* for 1 min at RT. Discard the upper ethidium bromide-containing phase.
19. Repeat the extraction until the DNA solutions have no color, and then perform one more extraction.
20. Dialyze the DNA overnight against three changes of dialysis buffer (1 L).
21. Transfer the DNA to microcentrifuge tubes, and ethanol-precipitate the DNA two times.
22. Resuspend the DNA in 200–400 μL of TE, and determine the DNA concentration by 260 nm absorbance.

## *3.5. Restriction Enzyme Mapping of Genomic DNA*

Restriction enzyme mapping is used to determine the size and orientation of the DNA inserts in positive clones and to provide preliminary information about the positions and sizes of introns. Precise intron locations are determined by sequence analysis. Our studies focused on the characterization of a clone containing a 13.5-kb genomic DNA insert (1A2.hg2).

1. Digest 1-μg aliquots of plasmid DNA with a number of restriction enzymes using 5 U enzyme/μg DNA.
2. Electrophorese the digests on a large-format 1% agarose gel in 1X TAE containing 0.1 μg/mL ethidium bromide. Photograph gel using a UV-transilluminator.
3. Incubate the gel in denaturing solution for 30 min at RT, in neutralizing solution for 30 min at RT, and then in 10X SSC for 10 min.
4. Transfer the DNA to Hybond-N nylon membrane by capillary action using a sandstone block in a reservoir of 10X SSC.
5. Prepare DNA probes as described in **Subheading 3.2.** The following probes were used in our analysis: a bovine cDNA obtained by degenerate oligonucleotide RT-PCR (1A2.bn1), a 1.6-kb *Eco*RI fragment of 1A2.hg2 containing the entire open reading frame, a 697-bp *Eco*RI-*Bgl*II fragment of 1A2.hg2 centered over the initiation codon, and an 827-bp *Pvu*II-*Pst*I fragment of 1A2.hg2 centered over the termination codon. Radiolabeled pGEM-5Zf(-) was used to identify plasmid bands.
6. Hybridize and wash the blot as described in **Subheading 3.3.1.**

## *3.6. Identification of the 5'-Untranslated Region Using RT-PCR*

RT coupled PCR is used to define the approximate size of the 5'-untranslated region when transcript abundance is too low to use techniques such as primer extension that can precisely locate the transcription start site *(5)*.

### *3.6.1. First-Strand cDNA Synthesis and PCR Amplifications*

1. Combine 100 ng human hippocampal poly(A) RNA, 0.8 m*M* oligonucleotide primer corresponding to the sequence upstream of the translation initiation site, and DEPC-treated water in a total volume of 16 μL. (We used primer [r2] corresponding to residues –151 through –169; (*see* **Fig. 1**).
2. Heat for 5 min at 70°C, and quick-chill on ice. Add 1 μL dNTP mix (25 m*M* each), 1.25 μL RNasin (50 U), 0.25 μL 100 m*M* DTT, and 5 μL 5X M-MLV reverse transcriptase buffer.
3. Add 275 U M-MLV reverse transcriptase, and incubate at 37°C for 1 h.
4. Heat at 95°C for 5 min to stop the reaction.
5. Prepare 50-μL reactions for PCR containing 1 μL cDNA, 200 μ*M* each dNTP, 1.5 m*M* MgCl$_2$, 1 μ*M* each upstream and downstream primer (*see* **Note 7**), and 2.5 U Taq polymerase.

```
                              (f1)
ATATGTGCAAAGCCTCCGAAGAGGATGGTTAAGTAAAGACTTAGGTTACCAGTATCAGGC   -1620
TTTCGTTTTTGTATGTAGGTAGCTCTACTGCCTCCTCTTAAAACCAACAAAGGAAAGAGA   -1560
GACTGGCTGCAAACTTTTAGAAGGAATGGCTTCGAATAGGGTTCCTGGGAGGAATCCCGA   -1500
                      (f2)
GGAAATAGACGCTGCTGCTCTGCTGATTGTCTCCACTATCCTGTTTTGCTCCTACCCACT   -1440
AATCCAGCCTGGGAGGCTCTGGGCATTAGCGGAAGGCTTCACCACAAGGAGACAGGAGCG   -1380
AGTATTCCATAGGCATGCGCTCCTAGTGGCACGAGTGGCTTGGGTCAGGATCAAAGAGTG   -1320
                                                  (f3)
AAGGATTCGGAAGTCAGCTATCTGGAGAGAGAGAGAGATTGTGTTTTATTCGTGTCCCAT   -1260
AGCTTTCCTATCCTATCCCTATCCTAGCTTTTAACCTGAGCCAGAGCTCACTACACAGGT   -1200
   (f4)
TCCTGGCTATCGAGTCTGAATCTGCACTACTCAACTTATAAACTGTCTGCAGACACCTGT   -1140
TAGGGAAATTGCTGATCATGGGCGGCAGGATCTGAACTCGCTTTACCTTCTTGTTTGGAG   -1080
CACAGGGACCGCCCAGCTAGAGGAGCACCAGCGCACTGCGCCCCAGCCCTGGGCGAGGGT   -1020
GCGGAGGATTTGTTCTCGGTGCAATCCTGCTGGCGCTTTTCCGGGGTTCTGCGCGGATCC   -960
(h1)
AGCTCCCCATCTCTGCTCCTACACACACAAAAGAAAACAACTCTCGATTGGAAGTTGTGG   -900
AATTTTCTCAGCCCCTACGAGGCGCGGGGATTCTCCAGCCCCGGCCCTCCTCCCGCCAGC   -840
CTGAGGTCTCCTTCGCTCGCCTGCCTTGCTAGGGACCGCAGTCCCTCAGCCGCAGCTGGG   -780
TCTGTCCGCCCCGCCTTTGCCCTCGCCTTTTCCCGGGGCGGATTTGGTGAAGTCGGCCTC   -720
                                                  (r1)
AAGTCCAGGAGGTCTGTCTTCGCCGGGCCAGCTCTCGCGGAACTGGGGGGTAGAGAGCAA   -660
AGGGAGAGATTCGTGGAAGGGAAGGGAGGTAGGGGTGGCGCAAACGCCCAGAGTATCAAA   -600
CTTGGGGGTGGCACAGTAGGTGACAGCAGCAGCTGCAGGTGGTGGCTGGGGACCCGCGAG   -540
GGGGCGCCCCTCTGGGTAGGGTCTGGCTGAGCGGGCTTGCAAGCCCGGGAGGCGGCTGAG   -480
AGACCCTGGACACTGTTCCTGCTCCCTCGCCACCAAAACTTCTCCTCCAGTCCCCTCCCC   -420
TGCAGGACCATCGCCCGCAGCCTCTGCACCTGTTTTCTTGTGTTTAAGGGTGGGGTTTGC   -360
CCCCCTCCCCACGCTCCCATCTCTGATCCTCCCACCTTCACCCGCCCACCCCGCGAGTGA   -300
GTGCGGTGCCCAGGCGCGCTTGGCCTGAGAGGTCGGCAGCAGACCCGGCAGCGCCAACCG   -240
CCCAGCCGCTCTGACTGCTCCGGCTGCCCGCCCGCGCGGCGCGGGCTGTCCTGGACCCTA   -180
   (r2)
GGAGGGGACGGAACCGGACTTGCCTTTGGGCACCTTCCAGGGCCCTCTCCAGGTCGGCTG   -120
GCTAATCATCGGACAGACGGACTGCACACATCTTGTTTCCGCGTCTCCGCAAAAACGCGA   -60
GGTCCAGGTCAgtaagtgtaattcctaactctgatcacctcctgggaacaagaaaatccc
tagaggagcgctcggttcctatgaagaggga.............................
............................................................
............... 4.5 kb intronic sequence ...................
.......................................................tttctttgctt
cacctttgtgtttttcctcgttccattggttttttgttgttgttgttttgttttatttttgtt
ttttcttttttagGTTGTAGACTCTTGTGCTGGTTGCAGGCCAAGTGGACCTGTACTGAAA   -1
ATGGGTCCAATAGGTGCAGAGGCTGATGAGAACCAGACAGTGGAAGAAATGAAGGTGGAA   60
 M   G   P   I   G   A   E   A   D   E   N   Q   T   V   E   E   M   K   V   E
```

Fig. 1. The sequence of the region encompassing the 5'-untranslated region of the human Y2 receptor gene. Numbering is with respect to the initiation codon. Lower case letters correspond to the intronic sequence. The partial amino acid sequence is shown in single letter code below the nucleotide sequence and corresponds to the published cDNA sequence (*see* **refs. 6** and **7**).

6. Prepare positive controls by substituting 10 ng of human genomic DNA for template cDNA. Assemble these reactions after all other PCR tubes are closed.
7. Cycle the reactions in a thermocycler running the following program: 3 min at 95°C followed by 40 cycles of 1 min at 94°C, 2 min at 53°C, 2 min at 72°C.

### 3.6.2. Hybridization Analysis of PCR Products

1. End-label an oligonucleotide corresponding to a sequence present in all predicted PCR products (we used primer h1; *see* **Fig. 1**) in a reaction containing 3 µL of 50 ng/µL oligonucleotide, 11 µL water, 2 µL 10X PNK buffer, 4 µL [$\gamma$-$^{32}$P]ATP, and 1 mL PNK. Incubate at 37°C for 30 min.
2. Heat inactivate at 65°C for 10 min.
3. Check the efficiency of the end labeling by PEI-chromatography as described in **Subheading 3.2.** Incorporation should be 50% or greater.
4. Purify the labeled probe on Sephadex G-50 as described in **Subheading 3.2.**
5. Run aliquots of the PCR products from **Subheading 3.6.1.** on an 1X TAE-agarose gel, and transfer the DNA to Hybond-N as described above.
6. Incubate the blot in oligo prehybridization solution (1 mL/8 cm$^2$ of membrane) in a sealable plastic bag on a shaking incubator for 30 min at a temperature 15°C below the $T_m$ for the oligonucleotide (*see* **Note 8**).
7. Dilute the end-labeled primer in 1 mL of oligo prehybridization solution, and add it to the bag containing the blot.
8. Hybridize in a shaking incubator for 60 min, the temperature used in **step 6**.
9. Wash the blot twice in 300 mL 2X SSC and 0.1% SDS for 10 min at RT, and then once in 300 mL of 5X SSC and 0.1% SDS for 10 min at the $T_m$ of the probe.
10. Wrap the blot in plastic wrap, attach fluorescent markers, and expose to film. The results of a typical experiment are shown in **Fig. 2**.

## 3.7. Expression and Competition Binding Assays

Competition ligand binding assays using peptides that distinguish between type 1 and type 2 NPY receptors **(8)**, as well as other closely related neuropeptide receptors, are used to determine the ligand-binding characteristics of cloned sequences.

### 3.7.1. Expression of Recombinant Protein in HEK293 Cells

1. On day 1, passage confluent cultures of HEK293 cells grown in DMEM containing 10% fetal bovine serum and antibiotics. Plate approximately $5 \times 10^6$ cells/ 15-cm tissue culture dish.
2. On day 2, remove the medium and add to each dish 12 mL transfection mixture containing: 400 µg/mL DEAE-Dextran, 0.1 m$M$ chloroquine, 2 µg/mL NPY2R.pcDNA3 in DMEM containing 10% fetal bovine serum and antibiotics (*see* **Note 9**).
3. Incubate the cells at 37°C and 5% $CO_2$ for 2 h, remove the transfection mixture, wash the cells with 15 mL PBS, and then incubate with 10% DMSO in PBS for 90 s.

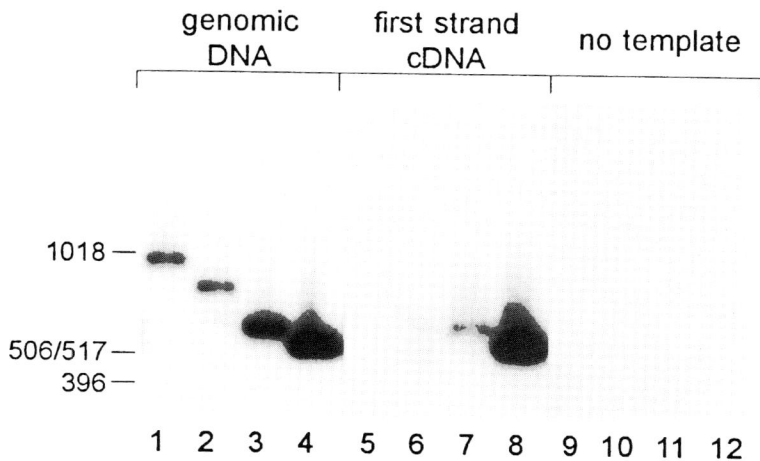

Fig. 2. Identification of the 5'-untranslated region of the human Y2 receptor gene. Human genomic DNA (lanes 1–4), first-strand cDNAs synthesized from human hippocampal poly(A) RNA (lanes 5–8), or no template controls (lanes 9–12) were subjected to PCR amplification using primers corresponding to Y2 receptor gene sequence upstream of the translation initiation site and intervening sequence. PCR of first-strand cDNA using primers corresponding to residues -1252 through -1270 (f3) and -1175 through -1192 (f4) (lanes 5 and 6) amplified products of the sizes expected from PCR amplification of genomic DNA. However, primers corresponding to residues -1630 through -1650 (f1) or -1464 through -1481 (f2) (lanes 7 and 8) did not reproducibly amplify the expected products. In contrast, all four products were amplified from reactions containing human genomic DNA (lanes 1–4), and were not seen in the absence of template (lanes 9–12). These results indicate that transcription begins within the region 200bp upstream of position –1270 (*see* **Fig. 1**).

4. Remove the DMSO solution, and wash the cells with 15 mL of DMEM containing 10% fetal bovine serum and antibiotics.
5. Add 25 mL of medium to each dish, and incubate at 37°C and 5% $CO_2$ overnight.
6. On day 3, change to fresh medium.
7. On day 4 (52–56 hours posttransfection), remove the medium, add ice-cold PBS, and scrape the cells from the dishes using a rubber policeman.
8. Transfer the cell suspensions to 50-mL centrifuge tubes on ice, rinse the plates with ice-cold PBS, and add the rinse to the cell suspension.
9. Centrifuge the suspensions at low speed in a clinical centrifuge, and then resuspend the cell pellets in ice-cold PBS.
10. Wash the cells by centrifugation two more times, and then resuspend the cells in ice-cold hypotonic solution containing protease inhibitors (2 mL/dish).
11. Quick-freeze the cell suspensions on dry ice, and store at –80°C until the day of the binding assay.

### 3.7.2. Membrane Preparation from Transfected Cells

1. Thaw cell suspensions in RT water, and transfer to a Teflon-glass homogenizer with a tight-fitting pestle attached to a variable-speed hand drill. Disrupt the cells using 40–50 up and down strokes with the homogenizer in ice. Monitor cell disruption by viewing drops of the cell suspension under a microscope at low power.
2. For each 12 mL of homogenate (from 5–6 plates), prepare a step-gradient containing 10 mL 40% sucrose solution overlaid with 15 mL 15% sucrose solution in an open top SW28 centrifuge tube.
3. Load 12 mL of homogenate on each gradient and centrifuge at 80,000$g$ in an SW28 rotor for 45 min at 4°C.
4. Recover the membrane protein band at the interface of the 15 and 40% sucrose solutions using an 18$^1$/$_2$-gage needle and syringe to pierce through the side of the tubes (*see* **Note 10**).
5. Dilute the band with 2.5 volumes of working buffer containing protease inhibitors.
6. Collect the membrane proteins by centrifuging at 200,000$g$ in a 70Ti rotor for 30 min at 4°C. Resuspend the pellets in membrane buffer (*see* **Note 11**), and determine protein concentrations using dye binding reagents.

### 3.7.3. Ligand Binding Assays

1. Prepare aliquots of peptides of interest (400 µL each) in a range of concentrations from $10^{-12}$ to $10^{-6}$ $M$ by serial dilution in binding buffer containing protease inhibitors. The concentrations of the aliquots should be 2.5-fold over that desired in the final assays.
2. Dilute $^{125}$I-Bolton-Hunter-labeled porcine NPY to 250 pM in binding buffer containing protease inhibitors.
3. Dilute HEK293 membranes to 3 mg/mL protein in binding buffer containing protease inhibitors.
4. Set up assays in 12 × 75-mm centrifuge tubes by adding the following solutions in order: 100 µL [$^{125}$I]NPY, 100 µL competitor peptide or buffer, and 50 µL membranes (100–150 µg protein). Prepare samples in triplicate.
5. Incubate samples with shaking for 1 h at RT.
6. Add 3 mL of ice-cold binding buffer to each sample, and filter through GF/C filters on a vacuum manifold.
7. Wash the GF/C filters two times with 3 mL of ice-cold binding buffer. Quantitate the radioactivity associated with membrane protein on the filters by γ-counting.
8. Binding data are analyzed using a program such as GraphPad. The results of a typical experiment are shown in **Fig. 3**.

## 4. Notes

1. Plates are incubated overnight at 37°C to remove excess moisture that interferes with the formation of discrete plaques. Fifteen-centimeter plates are used for primary screening, and 10-cm plates are used for second- and third-round screening.

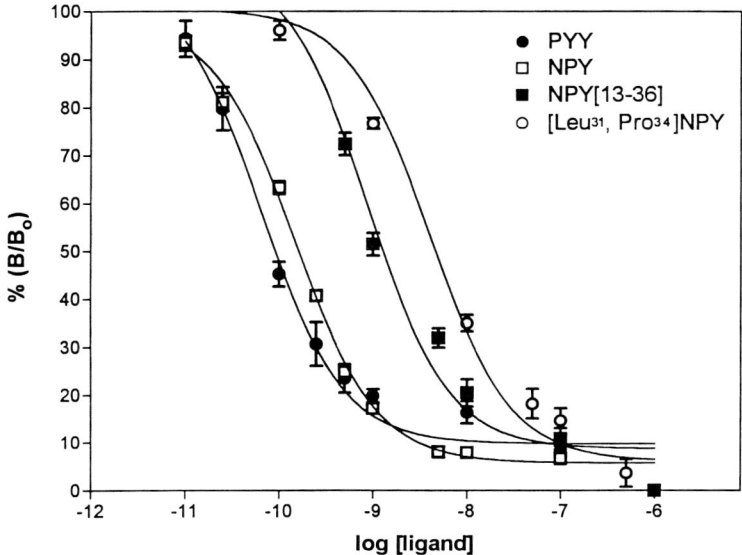

Fig. 3. Radioligand binding assay of recombinant human Y2 receptor. Competitions of $[^{125}I]$NPY binding were performed using NPY, PYY, $NPY_{13-36}$, and $[Leu^{31}, Pro^{34}]$NPY in the range of concentrations indicated in the figure. The binding of $[^{125}I]$NPY to membranes containing recombinant Y2 receptor exhibited a single class of high-affinity NPY binding sites ($K_D$ approx 0.5 n$M$), not present in membranes from untransfected cells. $[^{125}I]$NPY binding was not competed by 1 µ$M$ of substance K, substance P, neuromedin K, or the amino-terminal amidated peptides FLFQPQRF-NH$_2$ (F8-F-amide) and AGEGLSSPFWSLAAPQRF-NH$_2$ (A18-F-amide) (data not shown; *see* **ref. 9**). Competition binding of $^{125}I$-NPY with PYY and NPY$_{13-36}$ (IC$_{50}$ approx 0.1 and 0.9 n$M$, respectively) indicates that they bind recombinant Y2 receptor with affinities similar to that of NPY (IC$_{50}$ approx 0.2 n$M$). However, the $[Leu^{31}, Pro^{34}]$ substituted NPY analog was much less effective (IC$_{50}$ approx 4.0 n$M$). This type of binding is consistent with the type 2 NPY receptor subtype described by Wahlestedt et al. (*See* **ref. 8**.)

2. If available, prepare a positive control plate using phage containing NPY2R cDNA. Plate only about 50 pfu/10-cm dish, as large numbers of positives will deplete the radiolabeled probe from solution. Alternatively, use an NPY2R cDNA plasmid to transform *E. coli*, prepare plates with about 50 colonies/10-cm dish, and prepare lifts the same way as for phage plates.
3. Second lifts need to be left on the plates twice as long as the first to absorb the same amount of phage DNA. Mark the orientation of the second membrane using a needle without ink to pierce through at the position of the first hole seen as a light point on a black background.

4. Estimate the amount of cDNA recovered from low-melt agarose gels by running alongside a 1-µg aliquot of 1-kb DNA ladder (Gibco/BRL) on a 1X TAE-agarose gel. Compare the intensity of the sample band with that of the 1.6-kb standard band representing 100 ng of DNA.

5. Plaques contain roughly $10^7$ pfu/cm$^2$. Plugs obtained using the large end of glass pipets are approx 0.7 cm$^2$, and those from the small end are approx 0.02 cm$^2$. Therefore, phage stocks (in 0.5 mL SM) from primary plates contain roughly 10,000–15,000 pfu/µL, and phage stocks (in 0.1 mL SM) from secondary plates contain roughly 2000 pfu/µL.

6. Ligations are performed using a 3:1 molar ratio of insert to vector. In this case, "insert" includes not only a roughly 15-kb human genomic DNA insert, but also the 9- and 20-kb phage arms (approx 44 kb total), as the products from the restriction digest are not gel purified. The size of pGEM-5Zf(-) is approx 3 kb; therefore, a threefold molar ratio of insert to vector requires 44 µg of phage DNA/1 µg of plasmid DNA.

7. The downstream primer (r1) used in each PCR corresponded to residues -653 through –670 (**Fig. 1**). One of four upstream primers was used: (f1), -1630 through -1650; (f2), -1464 through -1481; (f3), -1252 through -1270; and (f4), -1175 through –1192 (**Fig. 1**).

8. $T_m$ is calculated according to the following formula: $T_m = 81.5°C + 16.6 \log [0.355] + 0.41 \times$ (GC content in%) – (600 ÷ oligonucleotide length).

9. A 1.6-kb fragment of a genomic clone (1A2.hg2) containing a 351-amino acid open reading frame encoding the type 2 NPY receptor was cloned into pcDNA3 (Invitrogen). *Exo*III and Mung bean nucleases were used to delete 394 bp of the 5'-upstream sequence (*10*) to create the expression plasmid NPY2R.pcDNA3.

10. Depending on the homogenization, one or two bands of membrane protein form on the gradient. The major band forms at the interface of the 15 and 40% sucrose solutions. A minor band sometimes forms at the interface of the homogenate and 15% sucrose solution. The binding properties of the two bands appear to be identical.

11. Work the membrane pellet into a homogeneous paste using a small glass or Teflon pestle, and then add membrane buffer a little at a time until the pellet is resuspended.

## Acknowledgments

Our work was supported by NIH grants EY09193 (to D.A.T.), EY07003 (Vision Core Grant), and RR00042 (U-M General Clinical Research Center).

## References

1. Buck, L. and Axel, R. (1991) A novel multigene family may encode odorant receptors: a molecular basis for odor recognition. *Cell* **65,** 175–187.
2. Larhammar, D., Blomqvist, A. G., Yee, F., Jazin, E., Yoo, H., and Wahlestedt, C. (1992) Cloning and functional expression of a human neuropeptide Y/peptide YY receptor of the Y1 type. *J. Biol. Chem.* **267,** 10,935–10,938.

3. Herzog, H., Hort, Y. J., Ball, H. J., Hayes, G., Shine, J., and Selbie, L. A. (1992) Cloned human neuropeptide Y receptor couples to two different second messenger systems. *Proc. Natl. Acad. Sci. USA* **89,** 5794–5798.

4. Ammar, D. A., Eadie, D. M., Wong, D. J., Kolakowski, L. F. Jr., Ma, Y.-Y., Yang-Feng, T. L., et al. (1996) Characterization of the human type 2 neuropeptide Y receptor gene and localization to the chromosome 4q region containing the type 1 neuropeptide Y receptor gene. *Genomics* **38,** 392–398.

5. Sambrook, J., Fritsch, E. F., and Maniatis, T. (1989) *Molecular Cloning: A Laboratory Manual*, 2nd ed. Cold Spring Harbor Laboratory, Cold Spring Harbor, NY.

6. Rose, P. M., Fernandes, P., Lynch, J. S., Frazier, S. T., Fisher, S. M., Kodukula, K., et al. (1995) Cloning and functional expression of a cDNA encoding a human type 2 neuropeptide Y receptor. *J. Biol. Chem.* **270,** 22,661–22,664.

7. Gerald, C., Walker, M. W., Vaysse, P. J.-J., He, C., Branchek, T. A., and Weinshank, R. L. (1995) Expression cloning and pharmacological characterization of a human hippocampal neuropeptide Y/peptide YY Y2 receptor subtype. *J. Biol. Chem.* **270,** 26,758–26,761.

8. Wahlestedt, C., Regunathan, S., and Reis, D. J. (1991) Identification of cultured cells selectively expressing Y1-, Y2-, or Y3-type receptors for neuropeptide Y/peptide YY. *Life Sci.* **50,** PL7-PL12.

9. Yang, H. Y., Fratta, W., Majane, E. A., and Costa, E. (1985) Isolation, sequencing, synthesis, and pharmacological characterization of two brain neuropeptides that modulate the action of morphine. *Proc. Natl. Acad. Sci. (USA)* **82,** 7757–7761.

10. Henikoff, S. (1987) Unidirectional digestion with exonuclease III in DNA sequence analysis. *Methods Enzymol.* **155,** 156–165.

# 4

# Y4 Receptor in Different Species

## Functional Expression and Binding

## Ingrid Lundell, Magnus M. Berglund, and Dan Larhammar

### 1. Introduction

The Y4-receptor subtype shows high species diversity compared with the Y1-, Y2-, and Y5-receptor subtypes. The rodent Y4 receptor differs considerably in sequence, pharmacology, and distribution from the human Y4 receptor. To characterize further the intriguing species differences of the Y4 receptor, we have also cloned the Y4 receptor from the guinea pig, which is evolutionarily nearly equidistant from both primates and rodents.

In previously reported studies on the cloned Y4 receptor, different binding properties and affinities for various ligands were reported *(1–8)*. Iodine-labeled peptide YY (PYY), [Leu$^{31}$,Pro$^{34}$]PYY, and pancreatic polypeptide (PP) have been used as radioligands in the various studies.

To be able to compare the pharmacological properties of the Y4 receptors from the three different species, we have carried out binding studies under the same laboratory setup using [$^{125}$I]hPP as radioligand, because this ligand has the highest affinity for the Y4 receptors. The coding regions of the receptor genes from human (h), rat (r), and guinea pig (gp) were generated by polymerase chain reaction (PCR) and cloned into the mammalian expression vector pTEJ-8 *(9)*. The receptors were stably expressed in Chinese hamster ovary (CHO) cells and assayed for [$^{125}$I]hPP binding as well as for the ability of various peptides and peptide analogs to displace radioligand binding *(2)*.

### 2. Materials

#### 2.1. Expression Constructs

1. Vent DNA polymerase (Biolabs, Beverly, MA).
2. QIAquick gel extraction kit (Qiagen, Hilden, Germany).

From: *Methods in Molecular Biology, vol. 153: Neuropeptide Y Protocols*
Edited by: A. Balasubramaniam © Humana Press Inc., Totowa, NJ

## 2.2. Expression of Receptors in Eukaryotic Cell Lines

1. LipofectIN (Life Technologies).
2. OptiMEM (Life Technologies, Paisley, UK).
3. Dulbecco's modified Eagle's medium (DMEM) (Life Technologies) with 10% fetal calf serum (FCS; Life Technologies).
4. Geneticin G418 (Sigma, St Louis, MO).
5. Trypsin-EDTA (Sigma).

## 2.3. Pharmacological Studies

1. 96-Well sample plate (polyethylene terephthalate [PET]; Wallac, Turku, Finland).
2. Binding buffer: 25 m$M$ HEPES buffer (adjust to pH 7.4 with HCl) 2.5 m$M$ CaCl$_2$, 1 m$M$ MgCl$_2$, and 2 g/L bacitracin (Sigma).
3. Wash buffer: 50 m$M$ Tris-HCl, pH 7.4, at 4°C.
4. Polyethylenimine 0.3% (Sigma).
5. [$^{125}$I]hPP; 2000 Ci/mmol (Eurodiagnostica, Malmö, Sweden).

## 2.4. Scintillation Counting

1. TOMTEC (Orange, CT) Mach III cell harvester.
2. Printed Filtermat A glass fibre filters (Wallac).
3. MeltiLex™ A Melt-on Scintillatior Sheets (Wallac).
4. 1495-021 Microsealer (Wallac).
5. 1450 Micro Beta Plus liquid scintillator counter (Wallac).

# 3. Methods

## 3.1. Expression Constructs

The coding regions of the receptor genes were cloned into the mammalian expression vector pTEJ-8. Two oligonucleotides were used as PCR primers to generate a fragment containing the entire coding region using the genomic phage clones (*see* **Note 1**) as templates. The PCR was run with Vent DNA polymerase (*see* **Note 2**). The forward and reverse primers, respectively, contained the sequence of two different restriction sites present in the expression vector to give directional cloning. The 5' primers were located a short distance upstream of the start codon, while the 3' primers were located a short distance downstream of the stop codon. All three constructs contained a *Hind*III cloning site in the forward primer, the human and rat constructs had an *Eco*RI cloning site in the reverse primer, and the guinea pig construct had a *Bam*HI site.

1. The 5' primer of the human Y4 receptor was CCG GGA AGC TTC CCG CGT CAT CCC TCA AGT GTA TC, and the 3' primer was CGG AAT TCC GGC AAG GGA CAT GGC AGG GAG.

2. The 5' primer of the rat Y4 receptor was CCG GGA AGC TTC CCT TTA GTC TTG AAG TTC CTG GTC T, and the 3' primer was CGG AAT TCC TGA AAG GGT GTG TCG AAA GAA AA.

3. The 5' primer of the guinea pig was CCG GGA AGC TTC ACT GCT GAC ATG GAC CGT TAG CTC, and the 3' primer was CGG GGA TCC GGG GAG AAG TCC TGG CAA GTG ACA T.

4. The PCR was carried out under the following conditions for the human and rat Y4 constructs: denaturation for 1 min at 94°C, annealing for 1 min at 50°C, and extension for 2 min at 72°C for 25 cycles, while the annealing temperature was at 45°C for the guinea pig construct, and the PCR was performed for 30 cycles.

5. The generated 1.25-kb fragments were phenol-extracted or gel-purified.

6. The PCR fragments and the expression vector pTEJ-8 were cut with the relevant restriction enzymes.

7. The fragments and vector were purified on an agarose gel using the QIAquick gel extraction kit.

8. The fragments were ligated into the expression vector and transformed into *Escherichia coli* cells.

9. Plasmid DNA was purified, and sequencing was performed to confirm the PCR-generated receptor constructs.

## 3.2. Expression of Receptors in Eukaryotic Cell Lines

### 3.2.1. Selection of Cells with Stable Expression

1. Grow CHO cells in 90-mm-diameter Petri dishes to a confluency of 40–50%.

2. Do a negative control plate without DNA in transfection reactions in parallel.

3. Linearize the expression constructs with *Pvu*I and phenol extract.

4. Mix 2 µg of the linearized DNA with OptiMEM in a total volume of 300 µL in a small Falcon tube.

5. Add 60 µL LipofectIN to 240 µL OptiMEM in another small Falcon tube.

6. Slowly drip the LipofectIN mix over the DNA solution while gently shaking.

7. Incubate in the Falcon tube at room temperature for 15 min.

8. Aspirate cell medium from cells and rinse with OptiMEM.

9. Add 4 mL of OptiMEM to the reaction, mix gently, and layer over the cells in the Petri dish.

10. Incubate cells for 6 h, and add 5 mL of DMEM containing 20% FCS.

11. Change media the following day to DMEM with 10% FCS.

12. Select for positive clones with 500 µg/mL of Geneticin (G418) in the growth medium for a period of 2–3 weeks until large clones bud off. Check negative control plate, which should have cells dying.

13. When isolated clones appear, pick positive clones (*see* **Note 3**).

14. Transfer single clones to separate wells in a 24-well plate. We recommend that you pick 48–72 clones.

15. After 2 days, trypsinate and resuspend cells carefully and transfer to 12-well plates.

16. When they are confluent, trypsinate and transfer clones to 6-well plates. In parallel, grow each cell line on a separate plate for freezing and storage.

### 3.2.2. Preparation of Cell Membranes

#### 3.2.2.1. SMALL-SCALE PREPARATION FOR TESTING OF EXPRESSION

The following steps need not be done in a sterile bench.
1. Rinse plate wells with phosphate-buffered saline (PBS).
2. Scrape off cells from plates and collect in 2-mL Eppendorf tubes on ice (*see* **Note 4**). Centrifuge and resuspend in PBS. Centrifuge again and resuspend in 500 µL binding buffer with bacitracin.
4. Homogenize cells on ice with an Ultra-Turrax homogenizer with a narrow rod to fit into the 2-mL Eppendorf tubes.
5. Test membrane preparations for binding.

#### 3.2.2.2. LARGE-SCALE PREPARATION FOR STORAGE

1. When a suitable clone has been selected (usually the clone with the highest expression), propagate this clone in several dishes to make a large-scale preparation.
2. After harvesting and final centrifugation, resuspend in binding buffer without bacitracin (*see* **Note 5**). Do not homogenize at this stage. Aliquot 100 µL into tubes for storage at –80°C.
3. To perform a binding assay, thaw an aliquot and place on ice.
4. Remove 10 µL and save for protein assay.
5. Dilute receptor membranes in binding buffer with bacitracin (*see* **Note 6**).
6. Homogenize membranes with an Ultra-Turrax homogenizer.

### 3.3. Pharmacological Studies

### 3.3.1. Testing of Stable Clones

Convenient total volume for binding assay is 100 µL (*see* **Note 7**). Perform assay for total binding in triplicate; assay in duplicate is sufficient for nonspecific binding.

1. Prepare a 400 n$M$ solution of unlabeled ligand for determination of nonspecific binding (*see* **Note 8**). For making serial dilutions, use binding buffer with bacitracin.
2. For each individual clone, add 25 µL of buffer each to three wells (wells 1–3) of a microtiter plate.
3. Add 25 µL of unlabeled ligand to the next two wells (wells 4 and 5).
4. Prepare radioligand sufficient for five wells for each clone (*see* **Note 10**). Add 25 µL radioligand each to wells 1–5.
5. Add 50 µL of receptor membrane preparation to wells 1–5 for each clone.
6. Shake plate gently and incubate at room temperature for 2 h.

### 3.3.2. Saturation Assays

Total volume in saturation assays can be 200 µL. Twelve concentrations of radioligand made by twofold serial dilutions are used.

1. Prepare 1.5 mL of a 400 n$M$ solution of unlabeled ligand for determination of nonspecific binding (*see* **Note 8**). For serial dilutions, use binding buffer with bacitracin.
2. For determination of total binding at each concentration: pipet 50 μL of buffer into each well of two duplicate rows of microtiter plates (wells 1–12 of rows A and B).
3. For determination of nonspecific binding at each concentration: pipet 50 μL of the ligand into each well of the next two duplicate rows of microtiter plates (wells 1–12 of rows C and D.)
4. Make 12 twofold serial dilutions of radioligand in separate Eppendorf tubes from a start concentration of 2 n$M$ of radioligand (*see* **Note 9**). Make 250 μL each of 12 serial dilutions.
5. Pipet 50 μL of each concentration of radioligand into quadruplicate wells of the microtiter plate (column 1, rows A–D) and so forth for all columns.
6. Aliquot 100 μL of homogenized receptor membranes to each well (columns 1–12, rows A–D).
7. Shake plate gently, and incubate at room temperature for 2 h.

### 3.3.3. Inhibition Assays

Convenient total volume for inhibition assays is 100 μL. Eleven concentrations of half-logarithmic dilutions of competing ligand are used.

1. Prepare 11 concentrations of inhibiting ligand in Eppendorf tubes: prepare 100 μL of the start concentration in tube 1. Add 100 μL buffer to tubes 2–11.
2. Titrate in 1/2 log steps: transfer 46 μL from tube 1 into tube 2. Mix gently and pipet 46 μL from tube 2 into tube 3, and so forth until tube 11.
3. Pipet 25 μL from each concentration of inhibiting ligand into duplicate wells (wells A and B) in each position (1–11) of the microtiter plate and so forth.
4. Repeat for each inhibiting ligand.
5. Determine total binding by adding 25 μL buffer to four wells at position 12.
6. Prepare 100 μL of a 400 n$M$ solution of unlabeled ligand for determination of nonspecific binding (*see* **Note 8**). Add 25 μL of this ligand to another four wells at position 12.
7. Aliquot 25 μL of radioligand into each well (*see* **Note 10**).
8. Aliquot 50 μL of homogenized receptor membranes to each well.
9. Shake plate gently, and incubate at room temperature for 2 h.

### 3.3.4. Scintillation Counting

1. Presoak the glass fiber filters in 0.3% polyethylenimine for at least 2 h.
2. Terminate reactions by rapid filtration in ice-cold wash buffer using a TOMTEC Mach III cell harvester, programmed for a filter wash of 5.0 mL.
3. Dry filters in an oven at 60°C (takes about 30 min).
4. Pipet an aliquot of radioactive ligand (for each concentration used), add directly onto filters, and let dry. This will be used to determine the actual amount of radioligand used.

5. Treat dried filters with MeltiLex™ A melt-on scintillator sheets, using a Microsealer apparatus.
6. Monitor radioactivity retained on the filters in a Wallac 1450 Betaplate counter.
7. Analyze results, for instance with the Prism 2.0 Software Package (Graphpad, San Diego, CA).

## 4. Notes

1. As the Y4 receptors do not contain an intron in the coding region, the genomic phage clone could be used as a template for the PCR reaction.
2. To generate the expression constructs, we used Vent DNA polymerase because this enzyme has a better proofreading ability than a regular Taq polymerase.
3. We have carried out the following procedure without using a sterile hood in our cell lab with no problems of airborne contamination. To pick isolated clones we have found that it is convenient to use a pipet with a 200-µL plugged tip. Under the microscope, select large positive clones and aspirate into a pipet tip. Place the contents in a well in a 24-well plate. Add growth media. The clone will originally stick to the plate and continue growing as a large budding clone. After 2 days, trypsinate and transfer to a plate with larger wells. The cells will now form a confluent layer upon growth.
4. Instead of using a commercial scraper, we use a rubber blade from a window cleaning scraper, which can be bought in most supermarkets. This rubber blade is very flexible and easy to handle and can be cut to a suitable size depending on the size of the well. It does not have to be sterile and can be reused between each clone after it is rinsed with water and dried with a paper towel.
5. Cells should be resuspended in binding buffer without bacitracin because bacitracin may interfere in the subsequent protein assay of membrane preparations.
6. The concentration of receptor membranes should be adjusted so that <10% of radioligand is bound.
7. Microtiter plates with 200-µL wells are used in all binding assays. In a typical binding assay, the following amounts would be used: 25 µL of nonlabeled peptide (or buffer), 25 µL of radioligand, and 50 µL of receptor membranes. For saturation assays, the volumes of each reagent are doubled.
8. For determination of nonspecific binding, guinea pig PP was used at a final concentration of 100 n$M$, but any ligand that binds with high affinity to receptor can be used.
9. The Y4 receptors have a very high affinity for the ligand PP. A final concentration of 0.5 n$M$ was generally sufficient to approach saturation of receptors.
10. The radioligand concentration in the inhibition assays should be approximately the same as the $K_d$ value.

## References

1. Bard, J. A., Walker, M. W., Branchek, T. A., and Weinshank, R. L. (1995) Cloning and functional expression of a human Y4 subtype receptor for pancreatic polypeptide, neuropeptide Y, and peptide YY. *J. Biol. Chem.* **270,** 26,762–26,765.

2. Eriksson, H., Berglund, M. M., Holmberg, S. K. S., Kahl, U., Gehlert, D., and Larhammar, D. (1998) The cloned guinea pig pancreatic polypeptide receptor Y4 resembles more the human Y4 than does the rat Y4. *Regul. Pept.* **75,76,** 29–37.
3. Gehlert, D. R., Schober, D. A., Gackenheimer, S. L., Beavers, L., Johnson, D., Gadski, R., Lundell, I., and Larhammar, D. (1997) [$^{125}$I]-Leu31-Pro34-PYY is a high affinity radioligand for rat PP1 and Y1 receptors: evidence for heterogeneity of pancreatic polypeptide receptors. *Peptides* **18,** 397–401.
4. Gehlert, D. R., Schober, D. A., Gackenheimer, S. L., Beavers, L., Johnson, D., Gadski, R., Lundey, I., and Larhammar, D. (1996) Characterization of the peptide binding requirements for the cloned human pancreatic polypeptide preferring (PP1) receptor. *Mol. Pharmacol.* **50,** 112–118.
5. Gregor, P., Millham, M. L., Feng, Y., DeCarr, L. B., McCaleb, M. L., and Cornfield, L. J. (1996) Cloning and characterization of a novel receptor to pancretic polypeptide, a member of the neuropeptide Y receptor family. *FEBS Lett.* **381,** 58–62.
6. Lundell, I., Blomqvist, A. G., Berglund, M. M., Schober, D. A., Johnson, D., Statnick, M. A., et al. (1995) Cloning of a human receptor of the NPY receptor family with high affinity for pancreatic polypeptide and peptide YY. *J. Biol. Chem.* **270,** 29,123–29,128.
7. Lundell, I., Statnick, M. A., Johnson, D., Schober, D. A., Starbäck, P., Gehlert, D. R., et al. (1996) The cloned rat pancreatic polypeptide receptor exhibits profound differences to the orthologous human receptor. *Proc. Natl. Acad. Sci. USA* **93,** 5111–5115.
8. Yan, H., Yang, J., Marasco, J., Yamaguchi, K., Brenner, S., Collins, F., et al. (1996) Cloning and functional expresison of cDNAs encoding human and rat pancreatic polypeptide receptors. *Proc. Natl. Acad. Sci. USA* **93,** 4661-4665.
9. Johansen, T. E., Schøller, M. S., Tolstoy, S., and Schwartz, T. W. (1990) Biosynthesis of peptide precursors and protease inhibitors using new constitutive and inducible eukaryotic expression vectors. *FEBS Lett.* **267,** 289–294.

# 5

# Neuropeptide Y Y5 Receptor Expression

## Eric M. Parker

## 1. Introduction

Neuropeptide Y (NPY) is one of the most abundant neuropeptides in the mammalian nervous system and is an important mediator of numerous physiological effects *(1)*. NPY elicits its manifold biological effects by interacting with at least six different G-protein-coupled receptors known as Y1, Y2, Y3, Y4, Y5, and y6 *(2)*. The NPY Y1 and Y5 receptors have recently been the subject of intense basic research and drug development efforts because these two receptors appear to mediate the effects of NPY on food intake and the regulation of body weight. Unfortunately, the study of the NPY Y1 and Y5 receptors has been hampered by difficulties in expressing these two receptors in cultured cells.

In this chapter, techniques for obtaining high-level expression of the NPY Y5 receptor are described. The first technique describes the use of the 5' untranslated region (5' UTR) of the human NPY Y5 receptor cDNA to improve receptor expression. It has recently been shown that while the coding region of the human NPY Y5 receptor gene is encoded on a single exon, the 5' UTR of the human NPY Y5 receptor gene is encoded by several exons *(3,4)*. Hence, a number of distinct human NPY Y5 receptor cDNAs that differ in their 5' UTR are produced via alternative splicing *(3,4)*. Interestingly, the nature of the 5' UTR has a significant impact on the level of expression of the human NPY Y5 receptor in cultured cells *(4)*. Methods for isolating various naturally occurring 5' UTRs of the human NPY Y5 receptor cDNA are described, and the impact of various 5' UTRs on the expression of the human NPY Y5 receptor is demonstrated. The second technique describes the construction of a hybrid receptor that incorporates sequences from both the rat and human NPY Y5 receptors. It was found that the rat NPY Y5 receptor could be expressed in cultured cells at

From: *Methods in Molecular Biology, vol. 153: Neuropeptide Y Protocols*
Edited by: A. Balasubramaniam © Humana Press Inc., Totowa, NJ

much higher levels than the human NPY Y5 receptor *(5)*. This prompted the construction of a hybrid rat/human NPY Y5 receptor that consisted almost entirely of a human NPY Y5 receptor coding sequence but was expressed at levels similar to that of the rat NPY Y5 receptor *(5)*. It is important to mention that these techniques have been found to be applicable to the expression of several other G-protein-coupled receptors, including the NPY Y1 receptor (Carol Babij and Eric Parker, unpublished observations).

## 2. Materials

### 2.1. Isolation of Various 5' Untranslated Regions of the Human NPY Y5 Receptor cDNA by 5' Rapid Amplification of cDNA Ends (5' RACE)

1. 5' RACE System for Rapid Amplification of cDNA Ends, version 2.0 kit (#18374-058, Life Technologies, Gaithersburg, MD), which includes diethyl pyrocarbonate (DEPC)-treated water, 10X polymerase chain reaction (PCR) Buffer (200 m$M$ Tris-HCl, pH 8.4, 500 m$M$ KCl), 25 m$M$ MgCl$_2$, 10 m$M$ dNTP mix (10 m$M$ each of dATP, dCTP, dGTP, and dTTP), 0.1 $M$ dithiothreitol (DTT), Superscript II reverse transcriptase (200 U/µL), RNase mix, binding solution (6 $M$ NaI), GlassMax spin cartridges, wash buffer, 5X tailing buffer (950 m$M$ Tris-HCl, pH 8.4, 125 m$M$ KCl, 7.5 m$M$ MgCl$_2$), 2 m$M$ dCTP, terminal deoxynucleotidyl transferase (TdT), 10 µ$M$ abridged anchor primer (5'-GGCCACGCGTCGACTAGACGGGIIGGGIIGGGIIG-3'), and sample recovery tubes.
2. Human total brain poly(A)$^+$ RNA (#6516-1, Clontech, Palo Alto, CA).
3. hY5 GSP1 primer (5'-CTGTAAGTCATCTACACTGC-3', corresponding to nucleotides 121–140 of the human NPY Y5 cDNA sequence [GenBank accession number U66275]) (*see* **Note 1**).
4. hY5 GSP2 primer (5'-GGCAGCAGTATTATTCTCTGTGGC-3', corresponding to nucleotides 57–80 of the human NPY Y5 receptor cDNA sequence).
5. hY5 GSP3 primer (5'-GTCTTGTTATAATACTCGTCGAGC-3', corresponding to nucleotides 29–52 of the human NPY Y5 receptor cDNA sequence).
6. 70% ethanol.
7. Sterile, distilled, and deionized water.
8. TE buffer: 10 m$M$ Tris-HCl, pH 8.0, 1 m$M$ EDTA.
9. TA Cloning Kit (#K2000-J10, InVitrogen, Carlsbad, CA).
10. Quick Change Site-Directed Mutagenesis Kit (#200518, Stratagene, La Jolla, CA).
11. Seamless Cloning Kit (#214400, Stratagene).

## 3. Methods

### 3.1. Isolation of Various 5' UTRs of the Human NPY Y5 Receptor cDNA by 5' RACE

Several alternatively spliced human NPY Y5 receptor cDNAs have been shown to exist that differ in the sequence of the 5' UTR *(3,4)* (**Fig. 1**). The various 5' UTRs of the human NPY Y5 receptor cDNA can be isolated by the following technique.

The 5' RACE System for Rapid Amplification of cDNA Ends, version 2.0 kit from Life Technologies is used for the isolation of the 5' UTRs of the human NPY Y5 receptor cDNA.

1. Mix 2 μg of human total brain poly(A)$^+$ RNA and 2.5 pmol of hY5 GSP1 primer in a 0.2-mL thin-walled tube, and bring the total volume to 15.5 μL with DEPC-treated H$_2$O. Incubate this mixture for 10 min at 70°C to denature the RNA, and then chill the mixture for 1 min on ice.
2. Add 2.5 μL of 10X PCR buffer, 2.5 μL of 25 m$M$ MgCl$_2$, 1 μL of 10 m$M$ dNTP mix, and 2.5 μL of 0.1 $M$ DTT. Incubate the mixture for 1 min at 42°C.
3. Add 1 μL (200 U) of SuperScript II reverse transcriptase, mix, and incubate for 50 min at 42°C to synthesize the first-strand cDNA.
4. Incubate the mixture at 70°C for 15 min to terminate the reaction.
5. Add 1 μL of RNase mix, and incubate for 30 min at 37°C to degrade RNA. If desired, the synthesized first-strand cDNA can be stored at –20°C.
6. Add 120 μL of binding solution (6 $M$ NaI) to the first-strand cDNA reaction.
7. Transfer the cDNA/NaI solution to a GlassMax spin cartridge, and spin at 13,000$g$ for 20 s to remove unincorporated dNTPs, hY5 GSP1 primer, and proteins.
8. Add 400 μL of cold wash buffer to the spin cartridge, and spin at 13,000$g$ for 20 s. Repeat this wash step three additional times.
9. Add 400 μL of cold 70% ethanol to the spin cartridge, and spin at 13,000g for 20 s. Repeat this wash step one additional time.
10. Transfer the spin cartridge insert into a fresh sample recovery tube. Add 50 μL of sterilized, distilled, deionized water (preheated to 65°C) to the spin cartridge. Centrifuge at 13,000$g$ for 20 s to elute the cDNA.
11. Mix 10 μL of the purified first-strand cDNA with 6.5 μL of DEPC-treated H$_2$O, 5 μL of 5X tailing buffer, and 2.5 μL of 2 m$M$ dCTP in a 0.2 mL tube. Incubate the mixture for 2 min at 94°C, and then chill the mixture on ice for 1 min. Briefly centrifuge the mixture in a microfuge to collect the contents in the bottom of the tube.
12. Add 1 μL TdT, mix gently, and incubate at 37°C for 10 min to tail the 3'-end of the cDNA with dC.
13. Heat-inactivate the TdT at 65°C for 10 min, then briefly centrifuge the mixture in a microfuge, and place on ice.
14. In a thin-walled tube on ice, mix 5 μL of 10X PCR buffer, 3 μL of 25 m$M$ MgCl$_2$, 1 μL of 10 m$M$ dNTP mix, 2 μL of 10 m$M$ hY5 GSP2 primer, 2 μL of 10 m$M$ abridged anchor primer, 5 μL of dC-tailed first-strand cDNA, 31.5 μL DEPC-treated H$_2$O, and 0.5 μL of Taq DNA polymerase (5 U/μL).
15. Perform the PCR with the following cycle conditions: 94°C for 1 min (1 cycle); 94°C for 30 s followed by 55°C for 30 s followed by 72°C for 2 min (35 cycles); 72°C for 5 min (1 cycle). Subsequently, hold the PCR reactions at 4°C.
16. Analyze 10 μL of the reaction by agarose gel electrophoresis.

```
hY5A   AGGGGTTCGAATTCCCCACCGCCGCCTCCAGTCCTGCTCCCGCTGCCGCCACCGCCGGGGTGCAG
hY5B   AGGGGTTCGAATTCCCCACCGCCGCCTCCAGTCCTGCTCCCGCTGCCGCCACCGCCGGGGTGCAG
hY5C   AGGGGTTCGAATTCCCCACCGCCGCCTCCAGTCCTGCTCCCGCTGCCGCCACCGCCGGGGTGCAG
hY5D   AGGGGTTCGAATTCCCCACCGCCGCCTCCAGTCCTGCTCCCGCTGCCGCCACCGCCGGGGTGCAG
hY5E   AGGGGTTCGAATTCCCCACCGCCGCCTCCAGTCCTGCTCCCGCTGCCGCCACCGCCGGGGTGCAG

hY5A   GAGCGATCGCGCTGGGCCGCGCTCCCGGGAGCCCAGGGCCTGCAGCGCCGGGGCCCCCGAGGTCTGCTCATTGTGT
hY5B   GAGCGATCGCGCTGGGCGCTGGGCC----------GCGCCCCGAGGTCTGCTCATTGTGT
hY5C   GAGCGATC----------GCGCCCCGAGGTCTGCTCATTGTGT
hY5D   GAGCGATCGCGCTGGGCCGCGCTCCCGGGAGCCCTGCAGGGCCTGCAGCGGCGGGGCGCGCCCGAG----------
hY5E   GAGCGATCGCGCTGGGCCGCGCTCCCGGGAGCCCTGCAGGGCCTGCAGCGGCGGGGCGCGCCCCGAG----------

hY5A   TTTTCAGGAAAAAGGAAGGGAAAGGGTGTTACAAGGAAAGGCTATCGGTAACAACTGACCTGCCACAAAGTTAGAAG
hY5B   TTTTCAGGAAAAAGGAAGGGAAAGGGTGTTACAAGGAAAGGCTATCGGTAACAACTGACCTGCCACAAAGTTAGAAG
hY5C   TTTTCAGGAAAAAGGAAGGGAAAGGGTGTTACAAGGAAAGGCTATCGGTAACAACTGACCTGCCACAAAGTTAGAAG
hY5D   ------GAAAAGGAAGGGAAAGGGTGTTACAAGGAAAGGCTATCGGTAACAACTGACCTGCCACAAAGTTAGAAG
hY5E   ------------------------------------

hY5A   AAAGGATTGATTCAAGAAA----------GACTATAATATGGAT
hY5B   AAAGGATTGATTCAAGAAAATGATGAAGTTACTCTGCCATGTATTGGGATGATTCCAAACAGGACTATAATAATATGGAT
hY5C   AAAGGATTGATTCAAGAAA----------GACTATAATATGGAT
hY5D   AAAGGATTGATTCAAGAAA----------GACTATAATATGGAT
hY5E   ----------GACTATAATATGGAT
```

17. Dilute the primary PCR reaction 1:5, 1:50, and 1:100 in TE buffer.
18. In a thin-walled tube on ice, mix 5 µL of diluted primary PCR reaction, 1 µL of 10 m$M$ hY5 GSP3 primer, 1 µL of 10 µ$M$ abridged anchor primer, 5 µL of 10X PCR buffer, 3 µL of 25 m$M$ MgCl$_2$, 1 µL of 10 m$M$ dNTP mix, 33.5 µL DEPC-treated H$_2$O, 0.5 µL of Taq DNA polymerase (5 U/µL).
19. Perform the PCR with the following cycle conditions: 94°C for 1 min (1 cycle); 94°C for 30 s followed by 55°C for 30 s followed by 72°C for 2 min (35 cycles); 72°C for 5 min (1 cycle). Subsequently, hold the PCR reactions at 4°C.
20. Analyze 10 µL of the reaction by agarose gel electrophoresis.

Fragments 150–300 bp should be observed in the primary PCR reaction and/or in the nested PCR reaction. These fragments can be subcloned into the pCR2.1 vector by means of the TA cloning kit and sequenced to confirm their identity. The sequence of the five distinct human NPY Y5 receptor 5' UTRs that have been isolated by this technique are shown in **Fig. 1** (**Note 2**).

The coding region of the human NPY Y5 receptor cDNA can be isolated as described by Hu et al. (*6*). The various 5' UTRs can be attached to the coding region of the human NPY Y5 receptor cDNA by various means (e.g., by use of the Seamless Cloning Kit). **Figure 2** demonstrates that addition of various 5' UTRs to the human NPY Y5 receptor coding region dramatically increases the expression of the receptor relative to the expression of the human NPY Y5 receptor coding region alone.

### 3.2. Construction of the Hybrid Rat/Human NPY Y5 Receptor

The rat and human NPY Y5 receptor cDNAs were cloned as described by Parker et al. (*5*) and inserted into the expression vector pcDNA3.1.

1. Digest pcDNA3.1-rY5 with the restriction enzyme *Mun*I. This generates a 1.47-kb fragment containing the rY5 5' untranslated region, the first 134 amino acids of the rY5 coding region, and part of the pcDNA3.1 expression vector.
2. Digest pcDNA3.1-hY5 with the restriction enzyme *Mun*I. This generates a 5.57-kb fragment containing amino acids 135–445 of the hY5 coding region and part of the pcDNA3.1 vector.
3. Purify the 1.47-kb fragment from the pcDNA3.1-rY5 *Mun*I digestion and the 5.57-kb fragment from the pcDNA3.1-hY5 *Mun*I digestion.
4. Dephosphorylate the 5.57-kb pcDNA3.1-hY5 fragment by treatment with alkaline phosphatase.

---

Fig. 1. *(previous page)* Nucleotide sequence alignment of the 5' UTR of the various human NPY Y5 receptor cDNA splice variants. The sequence of the five unique 5' UTRs that have been isolated by the technique described are shown (designated hY5A–hY5E). Nucleotides that are identical among the different splice variants are boxed. Dashes indicate gaps in the alignment. The ATG initiation codon is shown in bold.

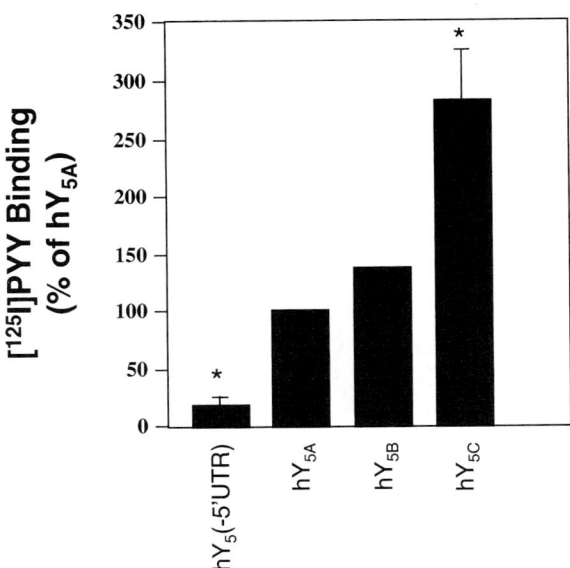

Fig. 2. Addition of 5' UTR to the human NPY Y5 receptor coding sequence increases the level of receptor expression. The human NPY Y5 receptor splice variants hY5A, hY5B, and hY5C (*see* **Fig. 1** for sequences of the 5' UTR in these constructs) and the human NPY Y5 receptor coding region without any attached 5' UTR (hY5(-5' UTR)) were transiently expressed in COS1 cells. The level of human NPY Y5 receptor expression in membranes from the transfected COS1 cells was determined by radioligand binding (0.1 n*M* [$^{125}$I]PYY), adjusted for transfection efficiency, and normalized to hY5A expression. The data shown are the mean ± SEM of three to four independent experiments. *, $p < 0.01$ compared with hY5A.

5. Ligate the 1.47-kb fragment from the pcDNA3.1-rY5 *Mun*I digestion to the 5.57-kb fragment from the pcDNA3.1-hY5 *Mun*I digestion, and transform the ligation products into competent *Escherichia coli*. Sequence the DNA prepared from the resulting colonies to confirm the sequence and orientation of the chimeric receptor.

6. Val$^{66}$ and Ala$^{112}$ of this construct (derived from the rat NPY Y5 receptor) can be changed to Leu$^{66}$ and Val$^{112}$, the corresponding residues in the human NPY Y5 receptor, by standard site-directed mutagenesis (e.g., using the Quick Change Site-Directed Mutagenesis kit).

The resulting chimeric NPY Y5 receptor consists of amino acids 1–35 of the rat NPY Y5 receptor and amino acids 36–445 of the human NPY Y5 receptor. A comparison of the levels of expression of the rat NPY Y5 receptor, the human NPY Y5 receptor, and the chimeric NPY Y5 receptor is shown in **Fig. 3**.

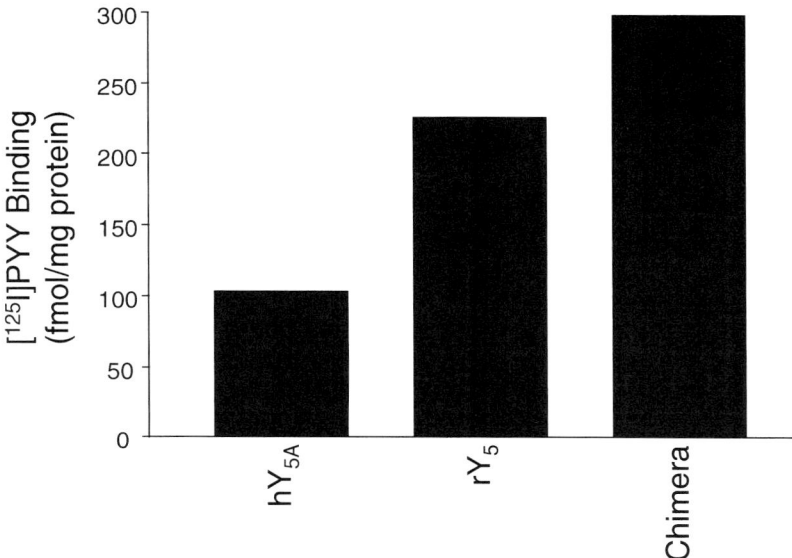

Fig. 3. The chimeric rat/human NPY Y5 receptor is expressed at high levels. The human NPY Y5 receptor splice variant $hY5_A$ (*see* **Fig. 1** for sequence of the 5' UTR of this variant), the rat NPY Y5 receptor (see **ref. 5** for sequence) and the chimeric rat/human NPY Y5 receptor prepared as described in this chapter were transiently expressed in COS1 cells. Membranes from the transfected COS1 cells were prepared, and the level of NPY Y5 receptor expression was determined by radioligand binding (0.1 n$M$ [$^{125}$I]PYY) and is expressed as fmol [$^{125}$I]PYY binding/mg protein. The data shown are from one experiment that is representative of three independent experiments.

Clearly, the chimeric receptor is expressed at levels similar to the native rat NPY Y5 receptor and at significantly higher levels than the native human NPY Y5A receptor.

## 4. Notes

1. The primers used in the first-strand cDNA synthesis and PCR reactions can correspond to other sequences in the NPY Y5 receptor cDNA. The specific primers listed have been used successfully, however, and should be used in initial efforts unless there are compelling reasons to do otherwise.
2. To rule out PCR artifacts in the preparation of human NPY Y5 receptor 5' untranslated region cDNAs, new primers having no overlap with the primers used in the 5' RACE reactions can be used in standard reverse transcriptase-PCR reactions with human brain poly(A)$^+$ RNA as template. Only human NPY Y5 receptor 5' untranslated region cDNAs that can be amplified with such independent primers are considered to be *bona fide* 5' RACE products.

## References

1. Wahlestedt, C. and Reis, D. J. (1993) Neuropeptide Y-related peptides and their receptors—are the receptors potential therapeutic targets? *Annu. Rev. Pharmacol. Toxicol.* **32,** 309–352.
2. Michel, M. C., Beck-Sickinger, A., Cox, H., Doods, H. N., Herzog, H., Larhammar, D., et al. (1998) XVI International Union of Pharmacology recommendations for the nomenclature of neuropeptide Y, peptide YY, and pancreatic polypeptide receptors. *Pharmacol. Rev.* **50,** 143–150.
3. Herzog, H., Darby, K., Ball, H., Hort, Y., Beck-Sickinger, A., and Shine, J. (1997) Overlapping gene structure of the human neuropeptide Y receptor subtypes Y1 and Y5 suggests coordinate transcriptional regulation. *Genomics* **41,** 315–319.
4. Parker, E. M. and Xia, L. (1999) Extensive alternative splicing in the 5' untranslated region of the rat and human neuropeptide Y $Y_5$ receptor genes regulates receptor expression. *J. Neurochem.* **73,** 913–920.
5. Parker, E. M., Babij, C. K., Balasubramaniam, A., Burrier, R. E., Guzzi, M., Hamud, F., et al. (1998) GR231118 (1229U91) and other analogues of the C-terminus of neuropeptide Y are potent neuropeptide Y $Y_1$ receptor antagonists and neuropeptide Y $Y_4$ receptor agonists. *Eur. J. Pharmacol.* **349,** 97–105
6. Hu, H., Bloomquist, B. T., Cornfield, L. J., DeCarr, L. B., Flores-Riveros, J. R., Friedman, L., et al. (1996) Identification of a novel hypothalamic neuropeptide Y receptor associated with feeding behavior. *J. Biol. Chem.* **271,** 26,315–26,319.

# 6

## Homology-Based Cloning Methods

*Identification of the NPY Y2, Y4, and Y6 Receptors*

**Douglas J. MacNeil and David H. Weinberg**

### 1. Introduction

Cloning strategies based on homology provide an attractive methodology for identifying novel members of a multigene family. The techniques are straightforward, with well-established protocols, and high-quality reagents are readily available. Most steps are relatively forgiving, and there is a short learning curve. The value of this should not be underestimated when comparing it with expression-based strategies, such as oocyte injections or transient cell transfections, in which success depends on meticulous replication of very exacting conditions. From a technical standpoint, homology-based approaches are the fastest and simplest to initiate and master. Of course, implicit in this is the assumption that novel family members will share significant homology to the known gene, which may or may not be true. Also, because the level of homology may be modest, low-stringency conditions are generally required, which results in a significant background of "false positives" in the assay. Indeed, a major challenge in the development of a homology-based approach lies in devising strategies that seek to maximize the efficiency of the process. In the course of our efforts we utilized four independent strategies, all of which led to the identification of novel genes. Each approach brought with it a set of advantages and disadvantages. Notably, no single strategy would have been sufficient to isolate all receptors.

### 2. Materials

Many of the techniques used in molecular biology today may be accomplished with all-inclusive kits that are constantly improving. In this chapter we focus on those components that are essential for successful implementation of

From: *Methods in Molecular Biology, vol. 153: Neuropeptide Y Protocols*
Edited by: A. Balasubramaniam © Humana Press Inc., Totowa, NJ

each strategy, the specific reagents we used are cited but should be taken as illustrative examples.

## 2.1. Low-Stringency Cosmid/Plasmid Library Screen

1. cDNA or genomic library from species of interest (*see* **Notes 1** and **2**).
2. 150 mm bacterial plates, minimum 84 for one library screen.
3. Standard reagents for gel electrophoresis, polymerase chain reaction (PCR), plasmid DNA preparation, and Southern blotting.
4. Hybridization solution: 5X standard saline citrate (SSC), 0.02% sodium dodecyl sulfate (SDS), 0.1% *n*-lauroyl sarcosine, 0.02% (w/v) blocking buffer (Boehringer Mannheim, Indianapolis, IN) (*see* **Note 3**).
5. Nylon hybridization membranes (e.g., NEF-978A, NEN Research Products, Boston, MA).
6. DNA fragment for radiolabeling and hybridization (*see* **Note 4**).
7. Kits for fragment labeling by random priming (e.g., Rediprime™, Amersham, Arlington Heights, IL) or PCR (e.g., PCR labeling kit, Life Technologies, Gaithersburg, MD) (*see* **Note 5**).

## 2.2. Multiple Probe Hybridization-Southern Blot Plasmid/Cosmid Library Screen

1. cDNA or genomic libraries from species of interest (*see* **Notes 1** and **2**).
2. Standard reagents for gel electrophoresis, PCR, plasmid DNA preparation, and Southern blotting. We prepared pool DNA using the Qiagen (Valencia, CA) midiprep system. DNA fragments for radiolabeling were purified from agarose gel using the Qiaquick (Valencia, CA) kit The antisense strand was labeled using the PCR labeling kit (Life Technologies).
3. Hybridization solution: Expresshyb™ hybridization solution (Clontech, Palo Alto, CA), which also contained 150 µL of 10 mg/mL sheared salmon sperm DNA solution (5 Prime-3 Prime, Boulder, CO) which had been heated to 95°C for 5 min, chilled on ice for 5 min, and then added to the hybridization solution. (*see* **Note 3**).
4. DNA fragment for radiolabeling and hybridization (*see* **Note 4**).
5. Kits for fragment labeling by random priming (e.g., Rediprime™, Amersham) or PCR (e.g., PCR labeling kit, Life Technologies) (*see* **Note 5**).
6. Wash buffer: 2X SSC, 0.1% SDS.

## 2.3 Degenerative PCR/colony hybridization screen

1. cDNA or genomic "target" DNA for PCR (*see* **Notes 1** and **2**).
2. Qiaquick (Qiagen) kit for purification of DNA from agarose gels.
3. PCR fragment cloning kits (e.g., T/A cloning kit, Invitrogen, Carlsbad, CA) for fragment cloning.
4. Hybridization buffers as described in **Subheadings 2.1.** or **2.2.** above and **Note 3**.
5. High-stringency wash buffer: 0.2X SSC, 0.1% SDS at 65 °C.

6. DNA fragment for radiolabeling and hybridization (*see* **Note 3**).
7. Kits for fragment labeling by random priming (e.g., Rediprime, Amersham) or PCR (e.g., PCR labeling kit, Life Technologies) (*see* **Note 5**).

## 2.4. Electronic Database Searching

1. Computer with Internet access.
2. Searching programs such as BLAST and TFASTA (Genetics, Madison, WI).

# 3. Methods

There are many excellent molecular biology protocol references available (e.g., **refs.** *1* and *2*). The researcher should be familiar with the general procedures for PCR, restriction enzymes, Southern blotting, and colony hybridization. In this chapter we focus on the specific details of each strategy that are paramount to its successful utilization.

## 3.1. Low-Stringency Hybridization to Plasmid/Cosmid Clones (Used to Clone NPY Y6)

At the time we first began to search for novel NPY subtypes, only NPY Y1 had been cloned *(3)*. We thus had no information to guide us to DNA sequence regions that might be conserved across family members. Therefore, our approach was to utilize a probe that spanned a large region of the gene (transmembrane domains 3–7) of the Y1 receptor and screen a cosmid library at relatively low stringency. Sequencing was then used to evaluate positives. Using this approach, we isolated the novel NPY receptor NPY Y6 *(4)*.

1. Place nylon filter membranes on 40 plates and allow to become wet for 1 min.
2. Plate cDNA or genomic library (*see* **Notes 1** and **2**) directly onto the membranes at a density of roughly 25,000 colonies/plate (*see* **Notes 6** and **7**).
3. Replicate membranes and process originals by standard protocols (i.e., denaturation, neutralization, and crosslinking).
4. Following DNA fixation to filters (e.g., by ultraviolet (UV) crosslinking), wash membranes in several changes of 1X SSC at room temperature using gloved fingers to rub off bacterial residue.
5. Wash membranes 30 min in 1X SSC, 0.1% SDS at 65 °C (*see* **Note 8**).
6. Prepare hybridization probe by random priming or PCR (*see* **Notes 4** and **5**).
7. Prehybridize membranes at 42°C for 1 h followed by addition of radiolabeled probe ($1.3 \times 10^6$ cpm/mL final activity) and hybridize at 42°C for 16 h.
8. Wash membranes in $2 \times 500$ mL of 1X SSC, 0.1 % SDS at room temperature, followed by $4 \times 500$ mL of 0.5X SSC, 0.1% SDS 55°C. (*see* **Note 9**).
9. Expose filters to X-ray film overnight (*see* **Notes 10** and **11**).
10. Select positives for further consideration, and repeat **steps 1–8**, using the bacterial colony corresponding to the hybridizing region and plating cells at higher dilution to obtain single pure clones.

11. Verify by direct DNA sequencing of plasmid DNA prepared from positive single clones.

## 3.2. Low-Stringency Southern Hybridization to DNA Derived from cDNA Pools in Plasmid Libraries (Used to Clone NPY Y2)

Since extensive colony hybridization at low stringency had only identified Y1 and Y6 clones (*see* **Subheading 3.1.**), we assumed that the homology between these receptors and the other NPY subtypes was going to be less than the 65% homology between the genes for Y1 and Y6. In our second round of screening, we instituted four improvements to our methodology. Pharmacological studies had determined that the neuroblastoma cell line SMS-MSN had Y2-like pharmacology *(5)*. Thus we chose to prepare a cDNA library from this cell line and screen it for Y2 clones. To help eliminate the background due to bacterial DNA and vector, we reduced the complexity of our target DNA and introduced a Southern hybridization-screening step. To help prioritize our "positives," we screened with multiple probes representing distinct NPY coding regions.

1. Plate $1 \times 10^6$ colonies on 150 filter membranes (100-mm-diameter) at a density of roughly 7000 colonies/plate (*see* **Subheading 3.1.** and **Notes 1, 2**, and **6**) and let colonies grow overnight at 37°C. Alternatively, 150 overnight cultures may be prepared, each inoculated with 7500 independent clones (*see* **Note 12**).
2. Add 5 mL of L broth to each membrane, and use a glass spreader to scrape off the bacteria and transfer to a disposable plastic screw-cap tube. Use 1 mL of each culture to prepare plasmid DNA. Adjust the remaining bacteria to 20% glycerol, quick freeze on dry ice, and store at –80°C.
3. Resulting DNA pools are subjected to restriction enzyme digestion to release the cDNA insert, and electrophorese through 1% agarose.
4. Photograph and process the gel, and transfer to nylon membrane for Southern blotting.
5. Prehybridize filters at 60°C for 1 h.
6. Add first labeled probe ($1.3 \times 10^6$ cpm/mL final activity), and hybridize at 60°C overnight.
7. Remove hybridization solution, rinse filters once in wash buffer at room temperature, and then incubate three times in 500 mL of wash buffer: 1 h at room temperature, 30 min at 30°C, and 20 min at 45°C.
8. Rinse again in room temperature wash buffer, wrap in plastic film, and expose to X-ray film overnight (*see* **Note 10**).
9. Adjust exposures as needed to ensure detection of bands other than vector.
10. Strip off residual radiolabeled probe by placing filters in 90°C water for 5 min.
11. Monitor with survey meter, and repeat if necessary. Expose the film overnight to confirm removal of radioactivity.
12. Repeat **steps 5–9** with a radiolabeled probe representing a separate sequence region.

13. Align X-ray films to identify pools containing bands hybridizing to both probes (*see* **Note 11**).
14. Use a small amount of the corresponding glycerol stocks to titer and plate at high dilution for colony hybridization. Isolate putative clones, and sequence inserts (*see* **Subheading 3.1.** and **Note 13**).

### 3.3. Degenerative PCR Based on Conserved Sequence Domains (Used to Clone NPY Y4)

Degenerative PCR has been used to isolate a number of G-protein-coupled receptors (GPCRs; e.g., *see* **ref. 6**). To select sequences for PCR primers, multiple receptor sequences are aligned, and the most conserved domains are used to design PCR primers. For GPCRs like the NPY receptors, these regions are usually the transmembrane domains. Based on the alignment of the Y1 and Y6 sequences, and later the Y1, Y6, and Y2 sequences, we designed a variety of primers (*see* **Table 1**), with degeneracy ranging from 1 to 512 (*see* **Notes 14 and 15**). These primers were then used pairwise against brain cDNA and human genomic DNA. We chose to use brain cDNA as a template because NPY receptors involved in feeding must be expressed in the brain (*see* **Note 1**). We chose to use genomic DNA as template because all NPY subtypes would be represented at the same molar ratio, and most GPCRs (including Y6, but not Y1) lack introns (*see* **Note 1**). Because this approach will result in the cloning of known sequences, we used subsequent hybridization to help eliminate them from consideration.

1. If multiple sequences are available, align sequences manually or with the aid of alignment software such as PILEUP (Genetics).
2. Design and synthesize PCR primers (*see* **Note 14**).
3. Assemble PCR reactions; total volume 50 μL with Taq polymerase using Taq buffer with 2.5 mM MgCl$_2$ (Life Technologies), 100 ng template DNA, and 50 pmol of each primer, using all combinations of primers.
4. Perform PCR. We used 35 cycles: denature at 95°C for 20 s, anneal at 45°C for 25 s, and extend at 72°C for 70 s (*see* **Note 16**).
5. Electrophorese reactions through 1% agarose, and isolate bands in the expected size range (*see* **Note 15**).
6. Purify DNA away from agarose.
7. Clone DNA fragments and transform *Escherichia coli*.
8. Plate on nylon membranes at moderate density (*see* **Subheading 3.1.**, 500/filter).
9. Process membranes for colony hybridization by standard methods (*see* **Subheading 3.1.**).
10. Hybridize with radiolabeled probes corresponding to known genes (e.g., Y1; *see* **Note 17**).
11. Wash membranes at high stringency (*see* **Subheading 2.3.**), and expose to X-ray film (*see* **Notes –11**) to identify exact match clones.

**Table 1**
**Primers Used for Degenerate PCR**

| Name | Oligo | Degeneracy |
|---|---|---|
| Forward | | |
| con-TM1 | GGA AAC CTI (G/T)CI (T/C)TI ATC AT(A/G/T) ATC AT | 12X |
| con-TM2a | CAT (C/T/A)CT GAT (C/T/A)G(T/C) IAA (T/C)CT ITC C(T/C)T CTC | 72X |
| con-TM2b | CAC TGG (G/A)TI TT(T/C) GGI (G/A)AI (G/A)CI ATG TG | 16X |
| con-TM2c | TC(C/A)T ICT IAT (C/T/A)G(T/C) IAA (T/C)CT I(T/G)C C(T/C)T CTC | 96X |
| con-TM2e | TACA C(A/C/T)I TI ATG GAC (T/C)AC TGG (A/G)T(A/C) TT(T/C) GG | 48X |
| con-TM3a | T(A/T/C)T TCT CIC TIG TI(C/T) TIA (T/C)(T/C)G CT(G/A) TIG AA | 48X |
| con-TM3b | CAG CTI AT(A/T/C) (G/A)TI AAC CCI (C/A)GI GGI TGG A | 12X |
| TM3c | TCA GCT IAT (C/A/T)(A/G)T IAA CCC I(A/C)(C/G) IGG ITG GA | 24X |
| TM2e | TACAC(C/A/T)ITIATGGAC(C/T)ACTGG(A/G)T(A/C)TT(C/T)GG | 48X |
| Reverse | | |
| con-CT | GT(A/G) TGC ATI GTI GA(G/C) ATG GC(A/T/G) AT | 12X |
| con-TM7a | GAA (G/A)TT TTT GTT IAG (G/A)AA ICC ATA A | 4X |
| con-TM7b | A(G/C)I A(G/C)(A/G) AAI AIC AG(A/G) T(T/C)G TGG T(A/T/G)G CAG | 96X |
| TM7c | TGAA(A/G)TTTTGTTIAG(A/G)AAICCATA(A/G)A | 8X |
| TM7d | TGT(A/G)GAI(A/G)(C/T)CAT(A/G/T)GCI(A/G)I(C/G)A(A/G)GTGGCA | 192X |
| con-TM6a | AGC AGI CII C(A/G)A AIG (T/C)IA CIA TGG A | 4X |
| con-TM6B | T(G/A)(G/A)G G(C/G)A(A/G) CCA G(C/A)(G/A) (G/A)CI (C/G)C(G/A) AA | 512X |
| con-TM6C | A(T/G) (A/G)(T/C)(T/C) T(C/G)I TG(G/A)T I CCA (G/A)TC | 256X |
| HELIX6.31 | GG AGC CAG CAG ACT GCA AAT GCT ACC ACA A | 1X |
| Y2 sequence included | | |
| TM2F | T ACC AACI TI(C/T)T I AT (C/T) G(C/T) (C/G) AA(C/T)CT | 32X |
| TM2G | TA(C/T) AC IIT (A/G) ATGG(A/G)(C/G)(C/G/T)A(C/G)TGG | 96X |
| TM3E | T GTI TCC A (C/T)(C/A/T)I TC(A/T)C I (C/T)TI (A/G)(C/T) | 96X |
| TM6D | T AG(A/G)G GCAG CCA G(A/C)I G (A/G)CI (C/G)C(A/G) AA | 32X |
| TM7F | T GTGGA (A/G)II CAT IGC I(A/G)IIAI GTGG(A/C)A | 8X |
| TM7G | T GG(A/G) TT(G/T) (A/G)(C/T)(A/G)(A/C)AI GTGGA | 128X |
| TM7H | T GAA (A/G)TT (G/T)ITGTT(C/G) A(G/T)I(A/C)A I CCATA | 32X |

*Method described in **Subheading 3.2.**

12. Isolate DNA from nonhybridizing clones, and sequence.
13. Identify abundant nonspecific products (*see* **Note 18**).
14. Use nonspecific product clones to generate radiolabeled probes and repeat **steps 9–11** (*see* **Note 19**).

## 3.4. DNA Sequence Database Searching: Identification and Evaluation of Orphan GPCRs as Putative NPY Subtypes

In addition to the homology-based experimental methods described above, electronic homology search methods provide a powerful tool in the search for new genes. Sequence repositories such as Genbank contain gene sequences, expressed sequence tags (ESTs), and raw genomic sequence. Using this procedure, we identified several orphans with significant homology to Y1, Y2, and or Y4. Expression studies subsequently demonstrated that these were not members of the NPY receptor family, but they have since been demonstrated to be neuropeptide receptors.

1. Select gene sequence(s) as probes for search (*see* **Note 20** and **Subheading 3.3., step 1**).
2. Perform search and evaluate "positives" (*see* **Note 21**).
3. Clone gene by PCR if full length or by the methods described in **Subheadings 3.1.–3.3.**

## 4. Notes

1. Both cDNA and genomic libraries are readily available from several commercial sources (e.g., Clontech and Stratagene [La Jolla, CA]). Genomic libraries have the advantage of a guaranteed representation of the entire genome of the organism. However, the abundance of nonexpressed sequence information contributes to nonspecific "false positives," which must be dealt with. Screening cDNA libraries can only be successful if the novel gene of interest is expressed in sufficient abundance so that it is likely to be represented in the number of clones evaluated. Ideally, both types of libraries are evaluated simultaneously. In addition, libraries are generated as inserts in either phage or plasmids/cosmids. For ease of manipulation, as well as for ease of signal detection, plasmid/cosmid libraries are preferable.
2. The success of homology-based approaches depends on similarity to known sequences, which can vary widely among species. In addition, some genes are only expressed within a limited species range (e.g., NPY Y6 is found in mouse but not in rat). Screening of multiple species is optimal.
3. The essential components of hybridization solutions are salt, pH buffering, and additives to prevent nonspecific binding. Within this framework there are many variations. The methods described in **Subheadings 3.1.** and **3.2.** utilize only two of these.
4. The length of the probe can greatly affect the success of the approach. Long probes provide larger regions of possible homology to the unknown target. How-

ever, this also greatly increases the likelihood of fortuitous nonspecific hybridization to uninteresting sequences, especially in the screening of genomic libraries. An alternative approach is to use two short probes representing two distinct regions and focus on positives identified by both independently (*see* **Subheading 3.2.**).

5. Probes may be prepared by several methods using commercially available kits. For probes around 500 nt long, we have found the best signal-to-noise ratios with PCR-generated probes. Specific activities of these PCR-derived probes also tend to be superior to that generated by other methods. For probes >500 nt and especially for longer probes, random priming has also worked well.

6. The titer on the filter membranes is usually lower than that found on plates and should be determined experimentally.

7. The number of plates needed depends on the total colonies screened and the density per plate. Colony density of $1 \times 10^6$ colonies is a reasonable starting number to screen for both cDNA and genomic libraries. About 25,000 colonies/plate allows colonies to grow to reasonable size, which yields better signals. Increased density allows use of fewer plates, but signals are harder to interpret.

8. Thorough washing of filters prior to hybridization greatly reduces nonspecific background. From this point on, filters are stable for several weeks wrapped in Saran Wrap at room temperature.

9. Washing is the most critical step in the process. Overly stringent washing (salt concentration too low, temperature too high) will result in the loss of weakly homologous clones. Too little washing will result in an unwieldy number of positives to sort through. The rate-limiting step is not in finding "positives" but in determining which ones are real (usually by sequencing). Since underwashed membranes can be rewashed without adverse effect (as long as they are kept under Saran Wrap so they don't dry out), the best strategy seems to be to wash under moderate conditions, determine the hit rate, and then rewash the filters at slightly higher stringency and repeat the evaluation. This approach allows positives to be ranked and prioritized for further evaluation.

10. Optimal exposure times are dependent on the quality of the radiolabeled probe, the sensitivity of the film, and the amount of DNA fixed to the membranes. This must be determined experimentally. An overnight exposure provides a useful reference point.

11. The number of "positives" identified is a function of hybridization/washing stringency (*see* **Note 9**) and exposure time. The number of clones one wishes to sort through is simply a personal preference. The use of multiple probes as outlined in **Subheading 3.2.** is one way to help weed out false positives.

12. Growing in culture medium is much less labor intensive than growing on plates. However, growth rate differences among clones are magnified in culture and may result in unequal representation of all clones.

13. Using this strategy, 53 clones were isolated that showed significant hybridization above background, and a single round of DNA sequence eliminated 49 of the clones as irrelevant. Complete sequences were determined from the remaining

four clones, one of which showed about 55% identity to the human Y1 sequence. At this time a patent application, WO 95/21245 from Synaptic Pharmaceutical Corp., was published and included the human Y2 receptor sequence that was a match for our clone (98.4% identical over 300 nt). Using the published sequence information, we then isolated a full-length NPY Y2 clone by PCR.

14. There are a variety of approaches to designing the primers: one can use primers with short highly conserved regions, long primers with high degeneracy, or primers with inosine as a nonspecific base.

15. In our experience, most primer pairs failed to generate fragments of the expected sizes, thus suggesting that it is important to test a variety of primers. Successful PCR reactions were obvious because they yielded bands 90 nt longer with genomic DNA than cDNA, due to the presence of an intron in the Y1 gene.

16. Optimal PCR conditions can vary from one thermocycler to another and should be determined with primers on a known target. The conditions described were used on a Omni Gene Thermocycler (National Labnet, Woodbridge, NJ).

17. Using various primers we were consistently able to isolate Y1 and Y6 sequences from genomic DNA, as well as Y1 sequences from brain cDNA. Of approximately 1600 colonies initially evaluated, more than one-third of the clones hybridized to Y1 or Y6.

18. DNA was isolated from 36 nonhybridizing colonies and sequenced as described above. Several clones contained Alu and other repetitive sequence elements and vector sequences from the cDNA libraries, but none of the clones had significant homology to the Y1, Y6, or Y2 receptors. Thus a second round of hybridization was performed using probes from the repetitive elements and non-GPCR sequences. These probes hybridized to another 700 clones.

19. After the second round, another 36 clones that had not hybridized to any of the probes were sequenced. One clone contained a sequence that was 66% identical to Y1 over 310 nt. This clone was determined to represent part of the Y4 receptor since it was 97% identical to the Y4 sequence in Synaptic patent WO 95/17906.

20. Test sequences for electronic searching may be DNA or protein. The redundancy of the genetic code and the evolutionary constraints on key amino acids for many peptides lead to an advantage for peptide sequence probes.

21. Electronic homology searches are completely equivalent to the experimental hybridizations described in **Subheadings 3.1.–3.3.** Low stringency (in this case percent overall identity) will yield enormous numbers of false-positive hits. High stringency will allow only the reidentification of sequences from which the probe was designed. As in those methods, probing with sequences from multiple regions and experimenting with different stringencies is important to the overall success of the method. We queried with the human Y1 peptide sequence using BLAST and TFASTA. We identified three orphans, the rat clone UHR1 *(7)*, the mouse clone mGIR, and the human clone GPR19 *(8)*, as well as three EST sequences (HG02, HG05, and HG31) with significant homology to Y1. None of these receptors promoted specific binding to radiolabeled NPY ligands ($[^{125}I]PYY$, $[^{125}I]PYY(3–36)$, $[^{125}I]Leu,ProNPY$, or $[^{125}I]rPP$; all available from NEN Prod-

ucts). Subsequent experiments have revealed that three of these receptors are neuropeptide receptors. HG02 and HG05 have been identified as orexin receptors *(9)*. UHR1 has been found to encode a receptor involved in prolactin release *(10)*.

## References

1. Harwood, A. J., ed. (1996) *Molecular Biology: Basic DNA and RNA Protocols*, vol. 58. Humana, Totowa, NJ.
2. Sambrook, J., Fritsch, E. F., and Maniatis, T. (1989*) Molecular Cloning: A Laboratory Manual.* Cold Spring Harbor Laboratory, Cold Spring Harbor, NY.
3. Larhammar, D., Blomquist, A. G., Yee, F., Jazin, E., Yoo, H., and Wahlestedt, C. (1992) Cloning and functional expression of a human neuropeptide Y/peptide YY receptor of the Y1-type. *J. Biol. Chem.* **267,** 10,935–10,938.
4. Weinberg, D. H, Sirinathsinghji, D. J. S., Tan, C. P., Shiao, L., Morin, N., Rigby, M. R., et al. (1996) Cloning and expression of a novel NPY receptor. *J. Biol. Chem.* **271,** 16,435–16,438.
5. Sheikh, S. P., Hakanson, R., and Schwartz, T. W. (1989) Y1 and Y2 receptors for neuropeptide Y. *FEBS Lett.* **245,** 209–214.
6. Lovenberg, T. W., Liaw, C. W., Grigoriadis, D. E., Clevenger, W., Chalmers. D. T., De Souza, E. B., et al. (1995) Cloning and characterization of a functionally distinct corticotropin-releasing factor receptor subtype from rat brain. *Proc. Natl. Acad. Sci. USA* **92,** 836–840.
7. Welch, S. K., O'Hara, B. F., Kilduff, T. S., and Heller, H. C. (1995) Sequence and tissue distribution of a candidate G-coupled receptor cloned from rat hypothalamus. *Biochem. Biophys. Res. Commun.* **209,** 606–613.
8. O'Dowd, B. F., Nguyen, T., Lynch, K. R., Kolakowski, L. F., Jr., Thompson, M., Cheng, R., et al. (1996) A novel gene codes for a putative G protein-coupled receptor with an abundant expression in brain. *FEBS Lett.* **394,** 325–329.
9. Sakurai, T., Amemiya, A., Ishii, M., Matsuzaki, I., Chemelli, R. M., Tanaka, H., et al. (1998) Orexins and orexin receptors: a family of hypothalamic neuropeptides and G protein-coupled receptors that regulate feeding behavior. *Cell* **92,** 573–585.
10. Hinuma, S. and Fukusumi, S. (1998) Ligand polypeptide for the G-protein-coupled-receptor from human pituitary glands. Patent WO 98/49295, published Nov. 5, 1998.

# II

## PRODUCTION OF TRANSGENIC AND KNOCK-OUT MODELS

# 7

# Developing Transgenic Neuropeptide Y Rats

## Mieczyslaw Michalkiewicz and Teresa Michalkiewicz

## 1. Introduction

Transgenic technology provides a powerful tool to study the function of a novel gene in the whole organism. The technology of transgenic rat production is receiving increasing attention. The rat is the model of choice in cardiovascular, endocrine, behavioral, and toxicology studies. This species offers a number of unique advantages for modeling human hypertension, inflammation, or neoplasia and has been used extensively for developing new therapeutic agents. The rat has also dominated in studies on the biology of NPY. Importantly, all of the known NPY receptors have been cloned, and novel NPY receptor antagonists have been designed based on the information generated in the rat. Therefore, NPY transgenic rats should provide useful models for studies elucidating the physiological role of NPY and designing candidate compounds for NPY receptor antagonists. Consequently, the accumulated wealth of published functional data from studies in rats makes interpretation of the results of transgenic studies in this species even more meaningful.

The notion that it is more difficult to generate a transgenic rat than a mouse stems from the historic fact that the mouse was initially used for transgenesis. As a result, straightforward application of the mouse transgenic protocol to the rat proved to be rather unsuccessful because of slight differences in the reproductive physiology of the rat. Although the efficiency of transgenic rat production is still generally lower than that of the mouse, the benefits of using the rat should outweigh the minor difficulties. Transgenic rats would bring more biodiversity to the field and would enable investigators to "walk" between these two species, thereby making comparative genomic and phenotypic analysis possible. It has been demonstrated that some genotypes can be better expressed in the rat than in the mouse. For instance, Mullins et al. (*1,2*)

From: *Methods in Molecular Biology, vol. 153: Neuropeptide Y Protocols*
Edited by: A. Balasubramaniam © Humana Press Inc., Totowa, NJ

have reported that development of the hypertensive renin transgenic rats was preceded by failure to develop human disease-like symptoms using the same transgene in the mouse.

Along with dramatic advances in the Human Genome Project, coordinating efforts are under way to develop important genomic tools for the rat genome. These include the genomic libraries, genetic map, expressed sequence tag, and sequencing of the rat genome *(3)*. These efforts will provide researchers with a number of novel genes whose function would need to be determined.

We developed a technique of transgenic rat production based on transgenic mouse technology. In this chapter we emphasize aspects of the transgenic technique specifically related to the rat and the NPY gene without providing many necessary details. Readers not familiar with the techniques of transgenic animal production are referred to the excellent available publications *(4–6)*.

## 1.1. Rat NPY Transgene: Design, Subcloning, and Preparation of DNA Samples for Microinjection

The design of the transgene is a critical element in a transgenic experiment. The major goal of this experiment was to use a gene addition mutation approach to determine the physiological role of NPY. Therefore our goal was to target the transgenic NPY expression to physiological sites and at the same time ensure that the transgene expression was subject to the same developmental and physiological regulation as its endogenous counterpart. To accomplish this, a 14.5-kb copy of the rat structural NPY gene *(7)* was used as a transgene. This transgene contains all the regulatory elements of the endogenous rat NPY gene: the rat NPY promoter and the NPY coding sequences with all exons and introns, as well as 5-kb and 1.5-kb fragments of the 5' and 3' regions, respectively. This design of the transgene should ensure that the transgenic NPY is expressed at the sites of native constitutive expression so that, for example, the excess NPY is secreted in the same synaptic clefts of the vascular sympathetic nerve terminals as the endogenous NPY.

The development of the NPY transgenic rats, bearing the rat structural NPY gene, was preceded by expression of a reporter gene hCMV-LacZ in early rat embryos and incorporation of two NPY chimeric NPY gene constructs, mouse metallothionein-human NPY cDNA, and human cytomegalovirus-hNPY cDNA in the rat genome. However, injection of those two subsequent heterologous DNA constructs did not produce a true founder. Nonetheless, during the course of that work we observed that 9-day-old embryos positive for the heterologous transgenes had a significantly higher NPY level in their whole-body homogenates. On this basis we concluded that an early and ambiguous expression of the NPY gene might be lethal to the developing embryos.

Therefore we focused our efforts on developing transgenic rats using a clone of the rat NPY genomic locus.

## 1.2. Production of Fertilized Eggs in Rats

For pronuclear injection to be efficient, it is important to obtain an ample amount of fertilized eggs at the desired time. Hormonally induced super-ovulation and synchronization of ovulation are used to achieve this. The choice of strain of the egg donor will determine the genetic background of the transgenics. Moreover, the strain can have a significant effect on the transgene expression and manifestation of the phenotype. We used Sprague-Dawley rats as egg donors since this strain of rats has been commonly used in cardiovascular and endocrine research including studies involving NPY.

## 1.3. DNA Injection into the Rat Pronucleus

Sufficient pronuclear injection is the key element in transgenic rat production. It requires mastering the technique of delivering of a minuscule amount of clean DNA precisely into the egg pronuclei as well as manufacturing miniature glass holding pipets and injection needles. Proper pronuclear injection is not practically possible without specialized training and equipment.

## 1.4. Production of Pseudopregnant Females and Implantation of Eggs into Oviduct of the Rat

It is important to use as embryo recipients females that are good foster mothers. The recipients should also respond consistently to injectable anesthetics and recover from the surgery quickly. We found that outbred Sprague-Dawley females are inexpensive and made good foster mothers. Implantation of embryos into the oviduct requires some basic knowledge of the rat anatomy and skills in sterile microsurgery.

## 1.5. Detection of the Transgenic NPY Sequences in the Offspring, Integrity of the Transgene, and Transgene Copy Number

The pups born from the eggs injected with NPY transgene are tested for the presence of the transgene. It is also important to evaluate intactness, copy number, and number of integration sites of the transgene. Since our transgene has an identical endogenous counterpart, we used a restriction fragment length polymorphism method to distinguish between the two. For this purpose, in the construction of the transgene, we added an *Eco*RI restriction enzyme site that is not present in the endogenous gene. It is then possible to identify transgenic founders by Southern blot analysis of tail DNA digested with *Eco*RI and hybridized to a probe, as shown in **Fig. 5**.

## 1.6. Breeding Scheme to Establish the Transgenic Line of NPY Transgenic Rats

Successful breeding is essential in establishing a transgenic line of rats, deriving of transgenic homozygotes, and generating sufficient numbers of transgenic animals to conduct experiments. Sufficient knowledge of rat reproductive physiology and understanding of mendelian genetics are important for this stage of work with transgenic animals.

## 1.7. Analysis of the Transgene Expression

Information on the magnitude and sites of transgene expression is important to determine a causal relationship between the phenotype and the transgene product. Ectopic expression sites of the transgene may confound interpretation of the results and understanding of the true physiological role of NPY. Since in this model the transgene is a copy of the rat structural NPY gene, we expected that the tissue distribution pattern of the transgenic NPY would overlap with its endogenous counterpart. This design of the transgene rendered the model more useful for study of the physiological role of NPY. However, for this reason we are not able to distinguish qualitatively between the transgene product and its endogenous counterpart. Therefore, we used a quantitative radioimmunoassay to compare tissue concentrations of NPY in transgenic and nontransgenic littermates. Quantitative immunocytochemistry and *in situ* hybridization will further characterize these mutant rats with regard to cellular distribution of the transgenic NPY.

## 2. Materials

## 2.1. Subcloning and Preparation of DNA Samples for Microinjection

1. Plasmid DNA, containing approximately 100 µg of the construct.
2. Restriction enzyme, *Not*I.
3. Agarose gel electrophoresis system.
4. Electro-Separation System (Schleicher & Schuell, Keene, NH).
5. Elutip Minicolumns (Schleicher & Schuell).
6. Ethanol.
7. Tris-EDTA buffer (TE): 10 m*M* Tris-HCl, 0.1 m*M* EDTA, pH 7.4, filter-sterile.

## 2.2. Production of Fertilized Eggs in Rats

1. Sexually immature (3–4 weeks-old) Sprague-Dawley female egg donors and mature stud Sprague-Dawley males (Hilltop Lab Animals, Scottdale, PA).
2. Pregnant mare's serum gonadotropin (PMSG, Sigma, St. Louis, MO) to increase the number of ovarian follicles.
3. Human chorionic gonadotropin (hCG, Sigma) to synchronize ovulation.

4. M-2 and M-16 media (Sigma) containing 30 mg/mL penicillin and 25 mg/mL streptomycin.
5. Embryo-tested sterile water (Sigma).
6. Hyaluronidase IV-S (Sigma).
7. 35-mm sterile tissue culture dishes (Corning, Corning, NY).
8. Pipet system for egg handling as described elsewhere *(4)*.
9. Paraffin oil (Fluka, Switzerland).
10. 5% $CO_2$ incubator.

## 2.3. DNA Injection into the Rat Pronucleus

1. Pronuclear injection system consisting of an inverted microscope with Nomarski differential interference contrast optics (Olympus, Tokyo, Japan), an air table (Micro-g, TMC, Woburn, MA), two mechanical micromanipulators (Leitz), a microsyringe for egg holding (model IM-5B, Narishigi, Tokyo, Japan), a microinjector (Eppendorf, Hamburg, Germany), a pipet puller (model P-87, Sutter, Novato, CA), a microphorge (Model, MF-9, Narishigi).
2. Microinjection chamber made of a microscope cover glass (**Fig. 2**).
3. Borosilicate glass capillaries (#TW 100-4, World Precision Instruments, Sarasota, FL) for the production of holding pipets.
4. Borosilicate glass capillaries with internal filament (World Precision Instruments, #TW 100F-4) for the injection needle.
5. Dental wax (Polysciences, Warrington, PA).

## 2.4. Production of Pseudopregnant Females and Implantation of Eggs into Oviduct of the Rat

1. Adult female Sprague-Dawley rats (Hilltop).
2. Vasectomized ACI male rats (Harlan Sprague-Dawley, Indianopolis, In).
3. Anesthetic, methohexital sodium (Brevital, Jones Medical Industries, St. Louis, MO).
4. Dissecting microscope (Leica) and surgical microscope (C. Zeiss, Germany).
5. Surgical scissors, surgical and ophthalmic forceps (model 2-111, Metico, Saratoga, FL).
6. Surgical silk and autoclip metal wound clipper.
7. Ophthalmic cautery (Roboz, Rockville, MD).
8. Hemorrhage sponges (Surgical Spears, Merocel, Mystic, CT).
9. Custom-made embryo transfer pipets (**Fig. 4**).

## 2.5. Detection of the Transgenic NPY Sequences in the Offspring, Integrity of the Transgene, and Transgene Copy Number

1. Proteinase K, phenol/chloroform, ethanol for DNA extraction.
2. Restrictrion enzyme, *Eco*RI.
3. Electrophoresis system.
4. Polaroid camera.

5. Nylon membranes.
6. DNA transfer system (Posiblot) and UV DNA linker (both from Stratagene, La Jolla, CA).
7. Polymerase chain reaction (PCR) machine (Robocycler, Stratagene), primers, and *AmpliTaq* DNA polymeraze (Perkin Elmer, Branchburg, NJ).
8. DNA hybridization and chemiluminescence detection system: psoralen-biotin, streptavidin, hybridization and washing buffers (Schleicher & Schuell).
9. Hybridization oven (#HB-1D, Techne, Duxford, UK).
10. X-ray film, Biomax MR (Kodak, Rochester, NY).

## 2.6. Analysis of the Transgene Expression

1. NPY antibody and radioimmunoassay (*see* **ref. 7** and **Note 1**).
2. Radioactively labeled NPY (Dupont, Wilmington, DE).
3. Pentobarbital (Sigma).

# 3. Method
## 3.1. Subcloning and Preparation of DNA Samples for Microinjection

1. 14.5-kb *Hae-24* clone of the rat NPY gene *(7)* is used to construct the transgene (**Fig. 1**). To make this clone useful for pronuclear injection, polylinkers containing *Not*I and *Eco*RI sites are added to each end of the molecule. The construct is then subcloned into the *Not*I site of a high-copy-number plasmid pGSII AK/BlscrII (Genosys Biotechnologies, The Woodlands, TX), a 3861-bp derivative of Bluescript II SK(+/–) (Stratagene).
2. The identity of the cloned fragment is then verified by restriction enzyme mapping, PCR amplification of several known fragments, and partial sequencing using primers located at three different locations of the known sequences of the gene (*see* **Note 2**).
3. Plasmid vector containing the NPY gene is digested with *Not*I restriction enzyme and fractionated using agarose gel electrophoresis.
4. The 14.5-kb DNA insert is easily distinguishable on the agarose gel from the 3.9-kb vector sequences and is excised from the agarose and electroeluted using an Electro-Separation System (Schleicher & Schuell).
5. The DNA is then further purified using Elutip Minicolumns (Schleicher & Schuell) followed by four to five washes in 70% ethanol (*see* **Note 3**).
6. The linear DNA is dissolved in TE buffer and its concentration and integrity is estimated by ultraviolet (UV) absorption spectrophotometry and agarose gel electrophoresis, respectively. The gene is injected at a concentration of 2 µg/mL (*see* **Note 4**).

## 3.2. Production of Fertilized Eggs in Rats

1. Sexually immature female egg donors are maintained on a constant 12-h light, 12-h dark cycle (lights on at 0700–1900 h) and are induced to superovulate by ip injection of 20 IU PMSG at noon (*see* **Note 5**).

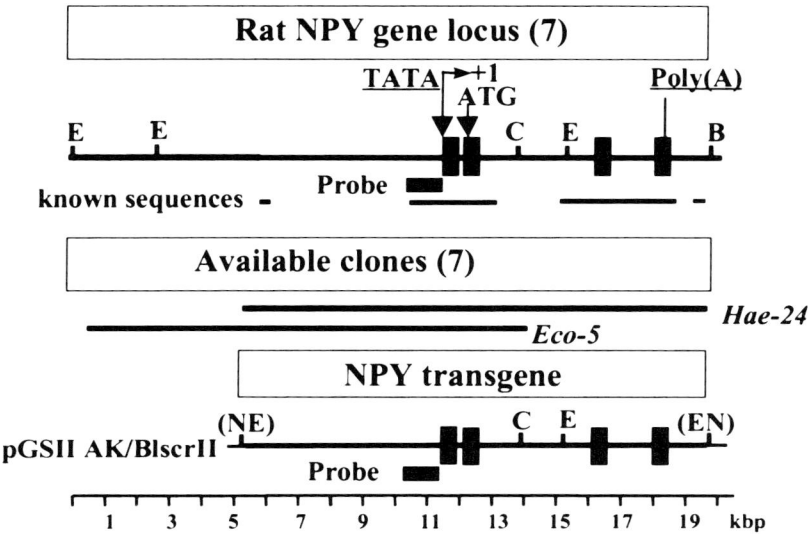

Fig. 1. NPY genomic locus in rat, available clones, and NPY transgene used for pronuclear injection. The known restriction sites are displayed *(7)*. The presumed TATA-like promoter element, transcription start, translation start, and poly-adenylation signals are marked as TATA, +1, ATG, and Poly(A), respectively. Known nucleotide sequences are marked. The transcription unit of the rat NPY gene is approximately 7.2 kb long and it is divided into four exons (vertical boxes). E, *Eco*RI; B, *Bgl*II; C, *Cla*I; (NE, EN), *Not*I; *Eco*RI (from polylinker). Thick horizontal bars represent hybridization probes.

2. Ovulations are synchronized by ip injection of 20 IU of hCG 48 h after PMSG injection. Immediately after hCG injection, each female is placed with a stud Sprague-Dawley male rat (*see* **Note 6**).
3. The next morning the females are checked for plugs, and those with plugs are killed between 10:00 and 11:00 AM by cervical dislocation *(4)* for egg collection.
4. The skin around the abdomen is sterilized with 70% ethanol. The abdomen is opened with a sterile scissors, and the oviducts are separated from the reproductive tract and placed in a culture dish containing M-2 medium (*see* **Note 7**). Under a dissecting microscope, the ampulla is incised with a sharp 19-gage needle, and the eggs are released into M-2 medium.
5. Using a mouth pipet (*see* **Note 8**), the eggs are then transferred to a 50–100 μL drop of M-2 containing 300 U/mL hyaluronidase and incubated for 10–20 min to digest the cumulus cells. To remove the enzyme and debris, the eggs are washed in M-2 by transferring them consecutively to 2–3 M-2 drops (*see* **Note 9**).
6. The eggs are then cultured for 2–3 h covered with paraffin oil M-16 medium in a humidified 5% $CO_2$ incubator at 37°C.

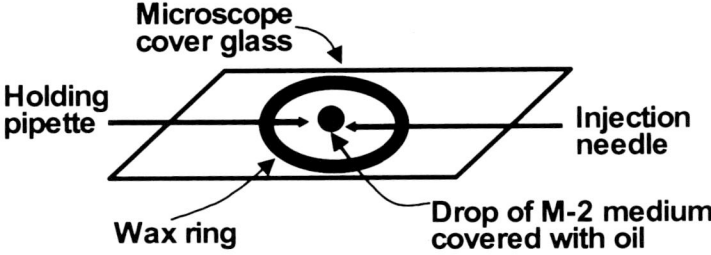

Fig. 2. Custom-made microinjection chamber. Hot dental wax is poured on a sterile microscope coverslip to make a ring shape. A drop of 10–20 μL of M-2 medium is laid within the ring, and the area is then covered with paraffin oil. The holding pipet is bent at a 25–30° angle in a flame to allow movement parallel to the surface of slide and better access to eggs

### 3.3. DNA Injection into the Rat Pronucleus

1. Borosilicate glass capillaries are used for making injection needles. The capillaries are pulled to have a long, narrow taper with a smaller opening than those recommended for the mouse. The needles are then "back-filled" with DNA solution by capillary action.
2. Holding pipets are produced as described elsewhere *(4)*.
3. For microinjection, the eggs are placed in an injection chamber made from a microscope cover glass (**Fig. 2**). Hot dental wax is poured in a ring-like shape on a sterile cover glass. A drop of 10–20 μL of M-2 medium is laid within the ring, and the area is then covered with paraffin oil.
4. Pronuclear injection is performed essentially as described for the mouse (*see* **ref. *4*** and **Note 10**). The egg is steadily positioned with a holding pipet, the injection needle is then inserted into the pronucleus and the DNA solution is injected until a distinct swelling of the pronucleus, is observed (*see* **Fig. 3** and **Note 11**).
5. After DNA injection, the embryos are transferred into M-16 medium for 1–2 h of incubation in a 5% $CO_2$ incubator (*see* **Note 12**).

### 3.4. Production of Pseudopregnant Females and Implantation of Eggs into Oviduct of the Rat

1. On the day prior to egg injection and transfer, 12–15 Sprague-Dawley females are mated in individual cages with vasectomized ACI (gray fur) males (*see* **Note 13**). Next morning the females are checked for plugs (*see* **Note 14**).
2. Pseudopregnant recipients are anesthetized with Brevital (50 mg/kg, ip), and the abdominal area is shaved with a clipper and wiped with 70% ethanol (*see* **Note 15**). The rat is covered with a surgical drape, and the sides are wiped with an antiseptic solution of 10% povidone iodine (*see* **Note 16**).
3. A small (<1 cm) transverse incision 1 cm down from the spinal cord at the level of the last rib is made with dissecting scissors through the skin and abdomi-

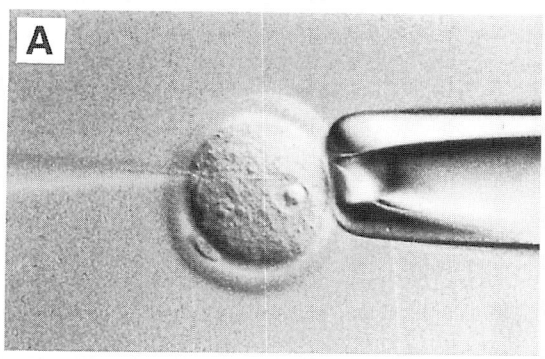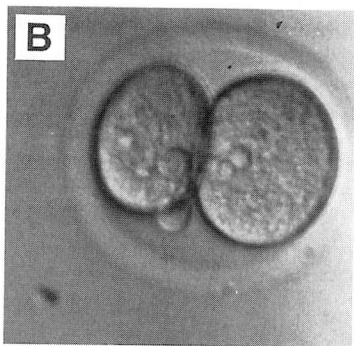

Fig. 3. DNA injection into a pronucleus of rat egg and two-cell rat embryo. (**A**) An egg is held by a glass pipet using gentle negative pressure from a Narishigi syringe (model IM-5B). The male pronucleus with the inserted injection needle containing the DNA solution is in focus. The female pronucleus is barely visible in this plane. After injection the pronucleus increases in diameter and becomes very sharp after smoothing out the membranes (magnification, ×250). (**B**) Properly injected and cultured eggs divide after overnight incubation.

nal muscles. The ovary is pulled out from the abdominal cavity and laid on sterile gauze.

4. Using a surgical microscope, the microvessels of the bursa are coagulated with an ophthalmic cautery. The bursa surrounding the ovary is then torn with dissecting forceps to expose the ostium of the oviduct. Eventual bleeding is stopped using a cautery or surgical sponge (*see* **Note 17**).
5. The infundibulum of the oviduct is picked up with fine ophthalmic forceps, an embryo transfer pipet (**Fig. 4**) with seven to eight injected embryos is inserted into the ostium (*see* **Note 18**), and the embryos are expelled into the infundibulum (*see* **Note 19**).
6. The reproductive tract is then replaced into the body cavity, and the muscles are sutured with silk thread followed by skin closure with wound clips. The area is wiped with an antiseptic solution of 10% povidone iodine. The procedure is repeated on the other side.
7. Postoperatively, the animals are placed on a heating pad and closely observed until they completely recover from anesthesia. The females are then kept in individual plastic cages in an animal facility (*see* **Note 20**).

## 3.5. Transgenic NPY Sequences in the Offspring, Integrity of the Transgene, and Transgene Copy Number

1. For DNA analysis, 2–3-mm tail biopsies are collected from pups at 2–3 weeks of age. At the same time the animals are marked using a toe-clipping system (*9*).

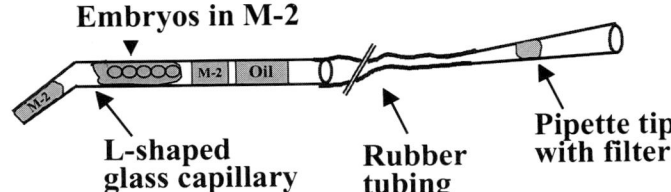

**Embryos in M-2**

**L-shaped
glass capillary**

**Rubber
tubing**

**Pipette tip
with filter**

Fig. 4. Embryo transfer pipet. Disposable 50-µL glass pipet is pulled in a flame and bent at the end at a 25–30° angle to allow easier penetration of the ostium. The wide end of the pipet is connected to rubber tubing, which is attached to a pipet tip with a filter. This disposable tip serves as a mouthpiece. First, paraffin oil is drawn up the pipet, followed by an air bubble, a drop of M-2 medium, and another drop of air bubble. Then 15–20 embryos dispersed in M-2 medium are sucked into the pipet. An air bubble and a drop of M-2 again follow the embryos. This combination of oil, air bubbles, and media in the transfer pipet provides greater controllability of embryo movement.

2. DNA is extracted after overnight tissue digestion with proteinase K using a standard DNA extraction protocol *(10)*.
3. DNA (15 µg) is digested with *Eco*RI, and fragments are separated using agarose gel electrophoresis. The DNA is then transferred onto a nylon membrane using the Posiblot DNA transfer system and UV crosslinked.
4. The hybridization probe (*see* **Note 21**), a 0.97-kb fragment of the rat NPY sequence, is amplified by PCR. Plasmid GSII AK/BlscrII containing the transgene (**Fig. 1**) is used as a template. The primer-binding sites flanked the fragment of the first exon and intron of the NPY gene (bp 377–1348 of the rat NPY sequence) *(11)*. PCR reactions (50 µL) contained 1 ng of supercoiled plasmid template, 250 n$M$ of each primer, 312.5 µ$M$ of each dNTP, 2.5 m$M$ of $MgCl_2$, and 1.25 U of *AmpliTaq* DNA polymerase. After an initial 5-min denaturation at 95°C, 35 cycles at 94°C for 60 s, 52°C for 30 s, and 72°C for 60 s are performed.
5. After purification and concentration of the PCR product using an ultrafilter (Centrex UF-2, Schleicher & Schuell), the probe is labeled with psoralen-biotin for chemiluminescence detection using Schleicher & Schuell kit.
6. The probe is hybridized for 16 h to Southern blots at 42°C with gentle rotation. After washing in 0.1X SSC, 0.1% SDS at 65°C, the membrane is incubated with blocking solution for 3 h and then with streptavidin-alkaline phosphate for 1 h. The blot is then placed on Lumi-Phos Substrate Sheet (Schleicher & Schuell) and exposed to X-ray film at 37°C (**Fig. 5**). Using this protocol, we were able to detect at least 1.5 pg of a 14-kb DNA fragment in Southern blot (*see* **Note 22**).

### 3.6. Breeding Scheme to Establish the Transgenic Line of NPY Transgenic Rats

1. Rat founder no. 400 carrying five copies of the transgene was mated with a nontransgenic partner (*see* **Fig. 6** and **Note 23**).

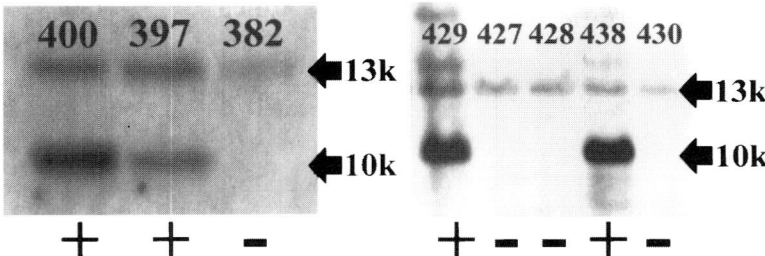

Fig. 5. Four transgenic rats carrying additional copies of the rat NPY transgene by Southern blot hybridization analysis. Rats born from the injected oocytes are screened for the transgene sequences by Southern blot hybridization using a 0.95-kb probe after digestion of tail DNA with *Eco*RI (**Fig. 1**). Numbers on the lanes represent rat identification numbers. Each lane shows the expected 13-kb signal from the endogenous NPY gene locus that is constant among all the offspring and the expected 10-kb signals from the NPY transgene. The hybridization signal from the endogenous NPY (13 kb) served as an internal control for estimation of the number of copies of the transgene. The 10 kb signals are the equivalent of the expected band from *Eco*RI digest of the transgene used for pronuclear injection (positive control, not shown). Rat nos. 400, 397, 426, and 438 are NPY transgenic founders with 5, 1, 10, and 24 copies of the transgene. Rat nos. 382, 427, 428, and 430 are nontransgenic littermates, with only the 13-kb band, indicating the presence of only endogenous NPY.

2. The progeny (F$_1$ generation) is then tested for the presence of the transgene as described earlier (*see* **Notes 24** and **25**).
3. To establish a line of NPY transgenic homozygotic rats, a heterozygote transgenic male is crossed with a hemizygotic transgenic female.
4. Homozygous rats are identified by Southern blotting as described earlier. The ratio of hybridization signals (quantified by scanning densitometry) between the endogenous gene and the transgene in the DNA from a candidate for homozygosity is about twice that from a hemizygous animal.
5. Homozygosity is confirmed by crossbreeding. A true homozygote, when crossed with a nontransgenic animal, will transmit the transgene to 100% of its offspring (*see* **Note 26**).

## 3.7. Analysis of the Transgene Expression

1. NPY transgenic rats from line no. 400 and their nontransgenic littermates are anesthetized with Pentobarbital (50 mg/kg, ip) and perfused with saline via the left ventricle.
2. For analysis of transgene expression at the protein level in the peripheral organs, a panel of peripheral tissues is dissected and quickly frozen.
3. The brain areas are micropunched from 300-μm sections under a dissecting microscope following an atlas of the rat brain (*12*).

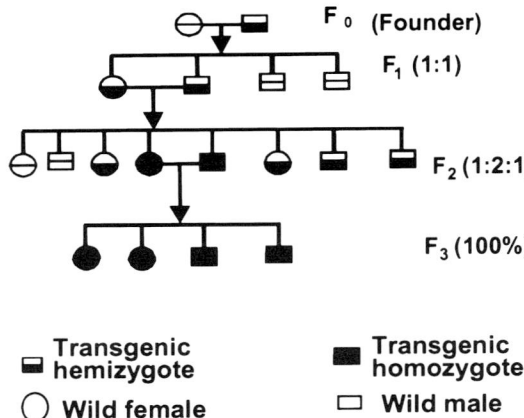

Fig. 6. Breeding scheme to establish transgenic line of NPY transgenic rats. Founder transgenic male $F_0$ is mated to a nontransgenic Sprague-Dawley female to generate a line of NPY transgenic rats. Approximately half (50.5%, $n = 109$) of the offspring from this mating ($F_1$ generation) was hemizygous for the transgene. Intercrosses between hemizygous rats within this line are set up to generate rats that are homozygous for the transgene. It is expected that approximately 25% of the offspring ($F_2$ generation) from this mating will be homozygous. Homozygosity will be confirmed by quantitative Southern blotting and confirmatory crossing with nontransgenic partners as described earlier. Numbers in parentheses indicate expected genotype ratios.

4. To determine whether the transgenic NPY is released after physiological stimulation, blood samples are collected from a chronic femoral vein catheter before and during 10 min of cold water stress.
5. NPY is then extracted from the tissues and assayed in triplicate using our specific radioimmunoassay (*see* **ref. 8** and **Note 27**).

### 3.8. Schedule and Efficiency of NPY Transgenic Rat Production

Transgenic rat experiments are very lengthy and consist of a number of different procedures; the results are never totally predictable. Therefore it is important to realize the time line and the difficulties of any particular stage of this technology (**Table 1**). It took a several months to construct the desired transgene that would address our hypothesis. Considering that in the rat, transgene incorporation generally occurs in only 5–10% of the offspring, the injection is performed for 3 weeks. During this time 501 embryos were injected and transferred to 33 recipients. Three weeks later, 108 live pups (no still birth was observed) were born. DNA analysis of tail biopsies revealed the presence of the transgene in six rats. The copy number of the transgene varied from 1 to

**Table 1**
**Schedule for Transgenic Rat Production and Phenotype Testing**[a]

| Time | Day | Procedure |
|---|---|---|
| Noon | −3 | Administration of PMSG into donors |
| Noon | −1 | Administration of hCG into donors. Placement of females with males |
| 7–9 AM | 0 | Checking for the copulatory plugs |
| 10–11 AM | | Collecting oocytes |
| 2–4 PM | | Pronuclear injection |
| 5–6 PM | | Transplantation of oocytes (501) into pseudopregnant females (33) |
| | 21 | Birth of $F_0$ generation (108) |
| | 49 | Weaning, DNA analysis, and selection of founders (6) |
| | 77 | Mating founders with normal mates |
| | 98 | Birth of $F_1$ generation |
| | 126 | Weaning, DNA analysis, and selection of transgenics (hemizygotes) |
| | 154 | Mating between $F_1$ transgenics |
| | | Expression analysis in $F_1$ |
| | | Beginning of phenotype testings in $F_1$ (hemizygotes) |
| | 175 | Birth of $F_2$ (homo- and hemizygotes) |
| | 203 | Weaning, DNA analysis and selection of transgenics |
| | 245 | Setting up crosses for testing homozygotes. |

[a]Number of animals is shown in parentheses.

24 in these founders. It is important to obtain and analyze a few founders. Each founder has different integration loci and copy number. The transgenic phenotype should be confirmed in other lines to ascertain that the observed effect is specific to the transgene product and not an artifact resulting from an insertional mutation, for example.

At age 10–11 weeks, the six founders were mated with wild partners (1) to provide sufficient animals to study the physiological effects of NPY overexpression; (2) to determine the transgene inheritance; and (3) to analyze the transgene expression pattern. It takes approximately 6 months after injection to generate a sufficient number of animals to start analysis of transgene expression and phenotype. To date, we observe that in all lines the NPY transgene is transferred to the next generation. A hypertensive phenotype is observed in four transgenic lines with more than three copies of the transgene but not in the two line bearers of one or two copies of the transgene (ms. submitted). This observation confirms that the hypertensive phenotype is specific to the transgene product and not an artifact due to the transgene integration site. Examination of the effect of NPY expression on other functions in these transgenic rats, including metabolic and behavioral functions, is in progress.

Generation of NPY transgenic homozygous animals is in progress. It takes at least 6–7 months to set up crosses to obtain homozygotes and another 4–5 months to set up crosses for testing homozygotes. Obtaining transgenic homozygotes allows comparison of transgene expression and the phenotypes with the hemizygous state (a gene dose-dependent effect). Assuming that homozygotes are viable, breeding to expand a line is much simpler, and it does not require genotyping because all offspring should inherit the transgene when homozygotes are bred together, with hemizygotes, or with nontransgenic mates.

## 4. Notes

1. The specificity of the radioimmunoassay is assessed using high-performance liquid chromatography and testing for crossreactivity with other members of the NPY peptide family *(11)*.
2. It is critically important to verify the correct cloning and the nucleotide sequence of the transgene before using it for pronuclear injection.
3. Purity of DNA is an important factor for microinjection efficiency, survival of the injected embryos, and transgenic production rate. Residues from purification columns may clog the injection needle. Contaminated water or alcohol residues in the DNA solution may drastically reduce embryo survival. Embryo-tested water (e.g., from Sigma) is recommended.
4. Since the vector fragments can diminish transgene expression and function, the transgene should be separated from the vector sequences for pronuclear injection; therefore, the construct should be designed so that these two components are easily separated from each other.
5. The young age of the donors is critical, as the older females will not respond to follicle-stimulating hormone. The strain of the egg donor may have an important effect on superovulation efficiency, visibility of the pronuclei, embryo survival, or transgene incorporation rate. We have not compared other rat strains as egg donors. Usually >90% of synchronized and superovulated Sprague-Dawley females mated, each giving 20–40 fertilized oocytes. Occasionally, females did not ovulate. Noise, crowding, or any other sudden disturbance in the room could affect superovulation or mating.
6. Hormones are prepared by dissolving the lyophilized powders in sterile saline to give a final concentration of 50 IU/mL. The hormones, in convenient aliquots, are then stored at –20°C for 1 year.
7. After adding the antibiotics, the media are filter-sterilized using a 0.22-mm cellulose filter and stored at 4°C for 1 month.
8. The pipet system for transferring the embryo is made of a heat-pulled glass capillary (50-μL disposable micropipet) attached to polyethylene tubing (ID 0.055 inches) and connected to a sterile pipet tip with a hydrophobic filter. A disposable pipet tip served as a mouthpiece.
9. Aliquots of hyaluronidase stock solution (20 mg/mL in sterile embryo-tested water) are stored at –20°C for 1 year.

10. More than 75% of the eggs were fertilized and had two visible pronuclei. Most of the pronuclei became visible in the afternoon. Therefore we started pronuclear injections after 2:00 PM. Compared with the mouse, the zona pellucida of the rat egg is considerably tougher, and the pronuclear membrane is more flexible. Because of this, it is harder to insert a pipet into the rat pronucleus. To break the durable zona pellucida and the flexible pronuclear membrane, the pipet is pushed much harder and farther into a rat egg than it is into a mouse egg. The flow rate is increased only when the needle tip is inside the pronucleus. A successfully injected rat pronucleus inflates dramatically, smoothing out the membranes, and the pronucleus becomes very sharply outlined.

11. It is usually necessary to readjust the focus for the next pronucleus.

12. Overnight incubation of the injected eggs makes it possible to evaluate whether the procedure of pronuclear injection is performed properly. Normally, >75% of the injected eggs enter the two-cell development stage after overnight incubation (**Fig. 3B**). This indicates that the eggs are properly harvested, injected, and cultured. Sudden shrinking of the eggs in the drop culture may indicate evaporation of the media or change in pH. An excessive number of lesioned eggs (becoming darker and shrinking away from the membrane cytoplasm) after injection probably occurs because the volume of injected DNA is too high or there is an extremely large opening in the injection needle.

13. It is important that the vasectomized males used to produce pseudopregnant females be indeed sterile. This ensures that the pups born after embryo transfer originated from the injected eggs and not from those of the foster mother. To provide such a control, a strain of males with gray fur (ACI rats) is used as vasectomized males. Thus, an eventual incompletely vasectomized and nonsterile male would be detected through the birth of pups with gray fur.

14. Each day, on average, at least one of five mating females was in natural estrus and had a plug.

15. Rats respond inconsistently to injectable anesthetics. We found that Brevital is a good anesthetic for Sprague-Dawley rats. Practically no female was lost during embryo transfer ($n = 175$).

16. Preparation of a recipient mother requires major survival surgery with opening of the abdominal cavity. All surgical instruments are sterilized. The surgery is performed on a clean table wiped with a disinfectant before and after use. A blue sterile pad covers the table, and the surgeon wears a surgical face mask, sterile gloves, and a gown.

17. The ovarian bursa in the rat is highly vascular. The disruption of it necessary for embryo transfer results in intense bleeding. This impedes the egg transfer considerably. Before disruption of the bursa, we found it very helpful to coagulate all the blood vessels of the bursa with electric cautery.

18. The end of the transfer pipet is bent in a flame at a 25–30° angle. The L-shape of the pipet is of great advantage in inserting the pipet into the oviduct ostium.

19. About seven to eight eggs are transferred to each oviduct. We recently compared single-oviduct transfer with that to both oviducts and found no difference in fertility rate. Single-oviduct embryo transfer is less stressful for the recipients.

20. In the NPY experiment, 501 eggs were transferred to 33 recipients, out of which 24 gave birth to 108 live pups. On average, combining all data up to date, one pup is born out of 2.8 transferred eggs (35.9%, $n = 2246$). Recipients recovered within a few days after surgery, and no mortality or symptoms of infection or inflammation have been observed.

21. The hybridization signal from the endogenous NPY (13 kb) serves as an internal control for the estimation of the number of copies of the transgene. In each sample the optical density of the hybridization band released from the endogenous gene is compared with the optical density of the transgene band. The optical density of the hybridization signals is measured using a computerized image analysis system (Bioscan Optimas). The density ratios for the rat nos. 400, 397, 429, and 438 were 5.1, 1.3, 8.6, and 23.6, respectively, indicating that these transgenic founders have approximately 5, 1, 9, and 24 copy numbers of the transgene, respectively (**Fig. 5**).

22. The NPY transgene was detected in 6 of 108 pups born (5.5%). Southern blot analysis of the DNA extracted from an NPY transgenic rat ($F_1$ generation of line 400) digested with *Cla*I restriction enzyme (a single-transgene cutter; **Fig. 1**) and hybridized with the same probe resulted in the expected 14.5-kb band (not shown). This indicates that in this line of transgenic rats, the transgene is intact and is incorporated as a cluster at one site in the rat genome.

23. Six lines of NPY transgenic rats with 1–24 copies of the transgene have been generated. For our study we selected founder no. 400 with five copies of the trangene. Transgenic animals with two to five copies of the transgene are optimal for most purposes. An unusually high copy number of transgene often affects the genetic stability of the transgenic locus and also results in low phenotypic expression *(5)*.

24. Approximately half (50.5%, $n = 109$) of this male's offspring carried the transgene into $F_1$ and $F_2$ generations. These transgenics reproduce normally, giving 12–15 pups/litter. Altogether these results indicate that male no. 400 is a true founder, with the transgene stably integrated at one site in its genome and that there is no embryonic death among the transgenics.

25. Nontransgenic littermates are used as controls in phenotype and transgene expression analyses. Utilization of nontransgenic littermates as controls significantly enhances the accuracy of observation. Siblings have a very similar genetic background and are growing in the same uterine and external environment as their transgenic mates.

26. Since, in these mutant rats, the NPY transgene is inherited following the mendelian principle, it is expected that the population of $F_2$ generation of the transgenic heterozygote crosses will on average consist of 25% nontransgenic animals, 25% homozygotes for the transgene, and approximately 50% transgenic heterozygotes.

27. A number of tissues from transgenic line no. 400 have significantly higher NPY levels. The highest levels of NPY overexpression were found in blood vessels, heart, salivary gland, spleen, lung, or hypothalamic nuclei (1.7–3.0-fold higher

compared with tissues from nontransgenic litter mates). Basal plasma levels of NPY in conscious NPY transgenic rats are not different, but cold stress-induced NPY release was significantly enhanced in NPY overexpressors. The expression pattern and NPY release in NPY transgenic rats indicate that this transgene is subjected to physiological regulation.

## Acknowledgments

We thank D. Larhammar and S. Lilleberg for the NPY clones and sub-cloning, respectively, D. L. Kreulen and H. Lubon for helpful discussions, and J. Steward and S. McDougal for editing the manuscript. This work was supported by grants from AHA, M. Puskar WVU, and NIH-NHLBI.

## References

1. Mullins, J. J., Sigmund, C. D., Kane-Haas, C., and Gross, K. W. (1989) Expression of the DBA/2J Ren-2 gene in the adrenal gland of transgenic mice. *EMBO J.* **8,** 4065–4072.
2. Mullins, J. J., Peters, J., Ganten, D. (1990) Fulminant hypertension in transgenic rats harbouring the mouse Ren-2 gene. *Nature* **344,** 541–544.
3. NIH Guide: Rat Genome Database. (1999) http://www. nih. gov/grants/guide/rfa-files/RFA-HL-99-013. html.
4. Hogan, B., Beddington, R., Constantini, F., Lacy, E. (1994) *Manipulating the Mouse Embryo.* Cold Spring Harbor Laboratory, Cold Spring Harbor, NY.
5. Pinkert, C. A. (ed.) (1994) *Transgenic Animal Technology.* Academic, New York.
6. Murphy, D., Carter, D. A. (eds.) (1993) *Transgenic Techniques.* Humana, Totowa, NJ.
7. Larhammar, D., Ericsson A., Persson H. (1987). Structure and expression of the rat NPY gene. *Proc. Natl. Acad. Sci. USA* **84,** 2068–2072.
8. Michalkiewicz, M., Huffman, L. J., Dey, M., and Hedge, G. A. (1993) Endogenous neuropeptide Y regulates thyroid blood flow. *Am. J. Phys.* **264,** E699–E705.
9. Gordon, J. W. (1993) Production of transgenic mice. *Methods Enzym.* **225,** 747–771.
10. Current Protocols in Molecular Biology (1994) Volume 1, Unit 2.2.
11. Genebank access No. M15792.
12. Paxinos, G. and Watson, C. (1982) *The Rat Brain in Stereotaxic Coordinates.* Academic, New York.
13. Michalkiewicz, M., Kreulen, D. L., Michalkiewicz, T., Lee, S. K., and Keith, R. L. Cardiovascular, metabolic and locomotor responses in neuropeptide Y transgenic rats. Submitted.

# 8

## Neuropeptide Y Y1 Receptor-Deficient Mice

*Generation and Characterization*

### Thierry Pedrazzini and Josiane Seydoux

### 1. Introduction

Energy homeostasis is the process by which adipose tissue, the stored body energy, is kept constant over time. Obviously, feeding behavior plays a crucial role in energy homeostasis. Therefore, to maintain a constant energy balance, energy (food) intake should match energy expenditure. A number of humoral factors are involved in signaling the body need in energy to modulate food intake. Among these hormones, leptin and insulin are secreted proportionally to the amount of body fat. They act in the central nervous system to regulate the action and expression of a variety of neurotransmitters and then modify energy balance by changing food intake and weight gain. The neurotransmitter neuropeptide Y (NPY) is a key player in the modulation of the leptin signal *(1)*. In particular, NPY promotes weight gain via its stimulatory effect on food intake in the hypothalamus. In addition, more evidence suggests that NPY also acts on specific target organs at the periphery to modify body fat stores. At least five different NPY receptor subtypes have been identified *(2)*. It has been hypothesized that each of these receptors might be associated with specific NPY functions.

To investigate the importance of the NPY Y1 receptors, we used a combination of molecular and physiological techniques to create and characterize mice that are deficient in the expression of this type of NPY receptor *(3)*. Indeed, in past years, the possibility of altering the mouse genome by introducing mutations at precise positions made possible the functional assessment of genes in vivo *(4)*. First, this technique, referred to as gene targeting in embryonic stem (ES) cells, takes advantage of our ability to culture ES cells in a pluripotent

From: *Methods in Molecular Biology, vol. 153: Neuropeptide Y Protocols*
Edited by: A. Balasubramaniam © Humana Press Inc., Totowa, NJ

state, i.e., retaining their capacity to colonize a normal embryo and to contribute to all developing tissues including the germline. Second, the technique makes use of the capacity of foreign genomic DNA to recombine with endogenous homologous sequences when introduced into the cells. Therefore, modified fragments of genomic DNA are used to create mutations at specific sites in the genome of ES cells. Then, these mutations can be transmitted to the germline of a chimeric mouse following transfer of the manipulated ES cells into developing embryos. In the case of the NPY Y1 receptor, gene targeting was used to introduce a null mutation in the Y1 locus to ablate the functions of the receptor *(3)*.

As mentioned above, NPY has dramatic effects on body fat mass. Therefore, as a first characterization of the induced phenotype, we determined body composition following complete desiccation and fat extraction. This technique allows precise measurement of body water, fat, and fat-free dry mass *(5)*.

## 2. Materials
### 2.1. Targeting Vector

1. Cloned gene from an isogenic library.
2. A good restriction map of the gene locus.
3. Plasmids containing selection markers (e.g., neomycin resistance gene, conferring resistance to G418, or thymidine kinase gene, conferring sensitivity to ganciclovir).

### 2.2. ES Cell Culture

1. ES cell line (we used HM-1 cells established in the laboratory of Dr. D. W. Melton, University of Edinburg, Scotland *(6)*; several other lines are available).
2. Fibroblast feeder cell line (e.g., SNL STO cells, not essential with HM-1 cells if using leukemia inhibitory factor (LIF).
3. High-glucose Dulbecco's modified Eagle's medium (DMEM); (GIBCO #41966-029).
4. 0.1% gelatin solution in $H_2O$ (Sigma #G-1890), autoclave.
5. Fetal calf serum (FCS), tested for supporting ES cell growth; inactivate at 56°C for 30 min.
6. Nonessential amino acids solution (100X, GIBCO #11140-035)
7. Sodium pyruvate (100X, GIBCO #11360-039).
8. L-glutamine (100X, GIBCO #25030-024).
9. Penicillin/streptomycin solution (100X, GIBCO #15140-114).
10. 0.1 *M* β-mercaptoethanol (1000X); prepare fresh by adding 100 mL of stock β-mercaptoethanol (Sigma #M-6250) to 14.1 mL $H_2O$, filter sterile, and store for up to 1 week at 4°C.
11. 0.025% trypsin/EDTA solution containing 0.1% chicken serum; mix 244.5 mL phosphate-buffered saline (PBS) with 0.5 mL 500 m*M* $Na_2$-EDTA, pH 8.0, 250 mL

trypsin/EDTA solution (0.05%, GIBCO #25300-062), 5 mL inactivated (56°C, 30 min) chicken serum (GIBCO #16110-033), filter sterile, and store frozen.
12. $10^7$ U/mL of LIF (Esgro, murine LIF, GIBCO #13275-029).

## 2.3. Electroporation

1. Hepes-PBS (HBS) (10X), dissolve 10 g HEPES 0.252 g $Na_2HPO_4·2H_2O$, 16 g NaCl, 0.74 g KCl, 2 g D-glucose in 180 mL $H_2O$, adjust pH to 7.2, and bring volume to 200 mL.
2. PBS 1X.
3. 200 to 300 µg linearized targeting vector in 100 µL TE (10 m$M$ Tris-HCl, 1 m$M$ EDTA, pH 7.4).
4. Electroporation system (Gene Pulser, BioRad #165-2098).
5. Electroporation cuvettes, 4 mm wide (BioRad #165-2088).

## 2.4. Selection

1. Geneticin (Sigma #G-9516, 100 mg/mL in 100 m$M$ HEPES, pH 7.2).
2. Ganciclovir (Syntex, Cytovene®, 2 m$M$ stock in $H_2O$).

## 2.5. Screening of Correctly Targeted ES Cell Clones

1. Polymerase chain reaction (PCR) primers for detection of the knockout mutation: forward: 5'-GTG GAT GTG.
2. GAA TGT GTG CGA G (spanning the pgk promoter present in the neomycin resistance cassette), reverse: 5'-TCT AGC CAG TTG GTA ATG GGT ATG G (homologous to a region of the Y1 gene immediately adjacent to the right arm of the targeting vector) (*see* **Fig. 1**).
3. PCR mixes:
   dNTPs (10 m$M$, Perkin Elmer #N808-0007).
   Taq2000™ DNA polymerase (5 U/mL, Stratagene #600196)
   10X reaction buffer (Stratagene)
   Lower mix: - dNTPs: 2 mL each
   Forward and reverse primers (50 m$M$): 1 µL each.
   $H_2O$: 15 µL.
   Upper mix:
   10X buffer: 10 µL.
   Taq2000™: 0.5 µL.
   $H_2O$: 39.5 µL.
4. DNA probe for Southern blot analysis (should be spanning a region outside the targeting vector)

## 2.6. Transfer of ES Cells into Mouse Blastocysts

1. Blastocysts from 3.5 d post coitus (p.c.) pregnant females.
2. ES cells in exponential growing phase.
3. Microscope with micromanipulators.

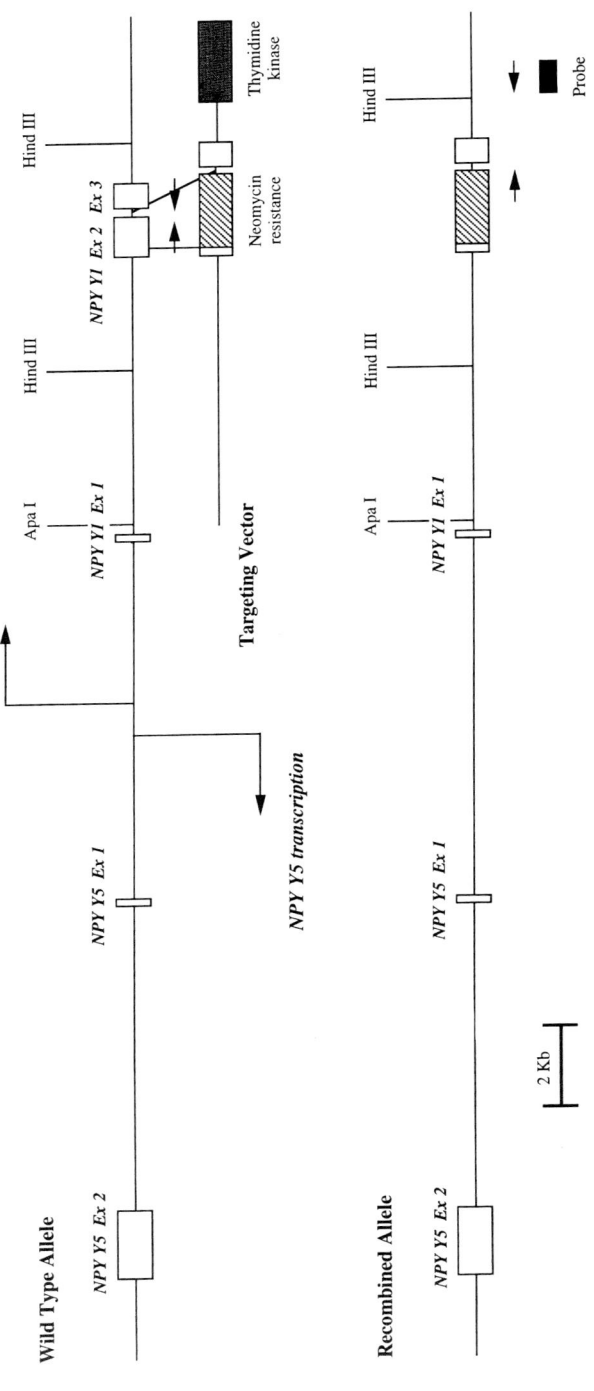

Fig. 1. The mouse NPY Y1/Y5 receptor genomic locus and targeting vector. The NPY Y1 and Y5 receptors are located in the same locus, where they are organized in an opposite orientation. The targeting vector used in Y1 gene disruption is presented as well as the resultant recombined allele. Arrows indicate positions of PCR primers used for the detection of either the mutated or the wild-type Y1 receptor. The dark box indicates the probe used in Southern analysis. This situation represents a good example of a complex gene organization, in which one wants to make sure that the mutation introduced affects one gene only. Indeed, the normal transcription of the adjacent Y5 gene had to be checked carefully.

4. Holding and injection glass pipets.
5. Pseudopregnant recipient females.

## 2.7. Screening and Breeding of Mice

1. Same as for identification of targeted ES cell clones.
2. PCR primers for the detection of the wild-type allele; forward: 5'-AAA TGT GTC ACT TGC GGC GTT C (spanning exon 2); reverse: 5'-TGG CTA TGG TCT CGT AGT CAT CGT C (spanning exon 3).

## 2.8. Body Composition

1. Oven.
2. Extraction thimbles (Schleicher & Schuelle #350219).
3. Pistil and mortar.
4. Homogenizer.
5. Petroleum benzine (Merck #UN1268).
6. Distillation apparatus.

# 3. Methods
## 3.1. Targeting Replacement Vector for the Y1 Locus

1. A targeting vector is defined as a fragment of genomic DNA, which has been modified to alter its sequences and is able to recombine and mutate a specific chromosomal locus. Primarily, gene targeting has been used to introduce null mutations in specific genes to assess gene functions. For this reason, replacement vectors are the most frequently used in gene targeting experiments. Standard molecular biology techniques are used to construct a targeting vector. These techniques will not be described here. However, **Fig. 1** depicts the replacement vector that we used to disrupt the NPY Y1 gene, and useful tricks are given in the notes regarding the design of targeting vectors (*see* **Note 1**).
2. The Y1 targeting vector was constructed around a 6-kb *Hind*III fragment isolated from a 129/J mouse genomic library using a rat probe spanning exon 2. This *Hind*III fragment contains the two coding exons (i.e., exons 2 and 3). A 0.8-kb *Eco*RI fragment containing most of exon 2 and part of the second intron was replaced by a neomycin resistance cassette. The left arm of the homology was made longer by adding a 4-kb *Apa*I/*Hind*III fragment spanning the region immediately adjacent to the 6-kb *Hind*III fragment. This results in a total length of homology of around 9 kb, with the left arm being 7.5 kb long and a right arm of 1.6 kb, ideal for PCR screening (**Fig. 1**).

## 3.2. ES Cell Culture

1. Prepare cell culture medium made of DMEM containing 15% FCS, L-glutamine, nonessential amino acids, sodium pyruvate, penicillin/streptomycin, and β-mercaptoethanol, with $10^3$ U/mL LIF.

2. Prepare gelatinized Petri dishes, i.e., cover the surface of the plates with 0.1% gelatin solution, leave for 20 min at room temperature, aspirate the solution, and let dry.
3. Thaw a frozen vial of HM-1 cells, resuspend in ES cell medium, wash twice in medium by centrifugation at 700$g$ for 10 min, and plate in 30-mm Petri dishes.
4. Change the medium the next day.
5. Passage the HM-1 cells on the second day, i.e., wash the cells with trypsin solution, add fresh trypsin solution, leave for 3–5 min at 37°C, add DMEM/15% FCS, pipet up and down, wash once, and plate on gelatinized dishes (0.5 × 10$^6$ cell/30-mm dish or 2 × 10$^6$ cells/60-mm dish).
6. Passage the cells every second day (*see* **Note 2** and **3**).

## 3.3. Electroporation

1. Trypsinize ES cells as above.
2. Wash once in PBS and twice in HBS 1X.
3. Adjust concentration at 70 × 10$^6$ cells/mL in HBS.
4. Mix 0.7 mL of cell suspension (50 × 10$^6$ cells) with 0.1 mL of DNA solution (300 μg of linearized plasmid) in a 0.4-cm electroporation cuvette.
5. Leave on ice for 10 min.
6. Electroporate at 800 V, 3 μF.
7. Leave cells on ice for 10 min.
8. Dilute cells with complete DMEM medium containing LIF at 10$^6$ cells/mL.
9. Plate on five gelatinized 100-mm Petri dishes, 10 mL/dish (*see* **Note 4**).

## 3.4. Selection

1. The day after electroporation, change medium (complete DMEM with LIF, no antibiotic selection).
2. The next day, change for medium containing Geneticin (350 μg/mL for HM-1 cells; the efficacy of this concentration should be checked if using a different cell line).
3. The next day, change for medium containing Geneticin and ganciclovir (2 μ*M*).
4. Change medium every day.
5. Resistant colonies should appear after 7–10 days.

## 3.5. Screening for Correctly Targeted ES Cell Clones

1. Replace medium with PBS.
2. Pick colonies using a 20-μL pipetman set to 5 μL (yellow tip).
3. Transfer each colony into a sterile Eppendorf tube containing 100 μL trypsin solution.
4. Incubate for up to 10 min.
5. Disperse each colony by pipetting up and down using a 200 mL pipetman (yellow tip), and transfer to a gelatinized 24-well plate containing 1 mL/well of complete medium with LIF (no selection).
6. Incubate for 2–4 days.

7. Replace medium with PBS.
8. Pick one colony from each well as above.
9. Transfer colonies in PCR reaction tubes (colonies can be pooled, 10 colonies/reaction tube), centrifuge, aspirate as much of the supernatant as possible, add 10 μL $H_2O$/tube, freeze on dry ice, and add 10 μL of 500 μg/mL proteinase K, and mineral oil. Incubate at 50°C for 60 min, inactivate at 95°C for 10 min, add PCR lower mix, centrifuge, denature (95°C/5 min), add upper mix, centrifuge, and perform PCR (usually 40 cycles: denature at 95°C for 1 min, anneal at 55°C for 1 min, and perform extension at 72°C for 3 min and final extension at 72°C for 10 min).
10. Repeat PCR analysis of individual colonies from positive pool.
11. Trypsinize each positive well, passage 1/10 of cells, and freeze the rest (freezing medium: DMEM, 20% FCS, 10% dimethylsulfoxide).
12. Culture the cells to confluence.
13. Passage each well into two 30-mm wells.
14. Culture the cells to confluence. Use one well for Southern analysis.
15. Purification of DNA for Southern analysis: wash twice in PBS, add 800 μL lysis buffer (50 m$M$ Tris-HCl, pH 8.0, 100 m$M$ EDTA, 100 m$M$ NaCl, 1% SDS), harvest in Eppendorf tubes, add 10 μL proteinase K (20 mg/mL), incubate overnight at 50°C, add 280 μL NaCl (5 $M$), vortex, centrifuge, transfer supernatants in new tubes, add 600 μL isopropanol, vortex, centrifuge, wash pellets in 70% ethanol, dry, resuspend in 100 μL TE, and use 25 μL for Southern.

## 3.6. Transfer of ES Cells into Mouse Blastocytes

The various steps are briefly described here. More detailed description can be found in several methodology books *(4)*.

1. Harvest blastocysts from a donor mouse.
2. Prepare a single cell suspension of ES cells.
3. Under the microscope, pick up about 20 ES cells into the injection pipet.
4. Maintain one blastocyst in position with the holding pipet, i.e., the blastocyst is oriented such that suction from the holding pipet is applied on the inner cell mass (ICM) side of the blastocysts.
5. Force the injection pipet through the trophectoderm, and inject ES cells as close as possible to the ICM (*see* **Note 5**).
6. Transfer approximately 15 injected blastocysts into the oviduct or the uterus of one 2.5 days p.c. pseudopregnant foster mother.

## 3.7. Screening and Breeding

Three different types of animals must be identified. First of all, chimeric mice can be recognized using coat color. This can also give useful indications on the degree of chimerism. For instance, chimerism is easily detected when ES cell lines from the agouti 129/SvJ strain are injected into blastocsyts from C57BL/6 mice with a black coat. Second, germline transmission of the chimeric animal is assessed by mating to mice with distinctive genetic markers. Coat color is again very convenient. The agouti progeny (129/SvJ × C57BL/6)F1 derived from 129/SvJ ES cells are readily identified from the black (C57BL/6 × C57BL/6)F1 puppies when chimeric parents are mated to C57BL/6 mice. Finally, one should be able to detect heterozygote and homozygote mutants from a general population.

1. Purify genomic DNA from tail biopsies.
2. Perform PCR analysis to detect null mutation, as was done for gene knockout ES cells (using the same primers, and approx 1 µg genomic DNA). Positive samples will identify heterozygote or homozygote mutants.
3. Perform PCR analysis to detect the presence of NPY Y1 exon 2 in samples, i.e., use the primer set described in **Subheading 2.7.** that is specific for exon 2, and the same PCR conditions. Positive samples will identify wild-type mice and heterozygote mutant animals.
4. By comparing the two PCR results, one can confirm the genotype of the transgenic mouse.
5. Alternatively, Southern analysis of genomic DNA can be used.

## 3.8. Body Composition

The method used is an adaptation of that described by Sohxlet *(5)*.

1. Weigh and sacrifice the mouse.
2. Incise the whole carcass, and dry to a constant weight in an oven at 60°C for 4–5 days. Calculate the water content by subtraction of dry weight from the whole body weight of the mouse.
3. Homogenize using a mortar and then a homogenizer to obtain a powder (*see* **Note 6**).
4. Place the entire homogenized sample (up to 7 g) in an extraction thimble and weigh.
5. Extract fat from the sample with distilled petroleum benzene, i.e., distill benzene by boiling at 60–80°C in a water bath, and refrigerate to condense benzene and ensure a constant dripping in the extraction thimble. After passing through the thimble, petroleum benzene with fat is returned to the distillation jar.
6. After 5–6 hours (until passing-through solvent is colorless), dry thimble in an oven.
7. Weigh the dry thimble. The difference between this weight and weight prior to fat extraction (*see* **Step 5**) equals to the total fat content of the animal.
8. Fat-free dry mass is obtained by subtracting body fat from the dry weight obtained in **step 2**.
9. Results of the analysis of NPY Y1-deficient mice are described in **Fig. 2**.

## 4. Notes

1. The design of a targeting vector should follow some rules, which are listed below:
   a. Screen a genomic library containing isogenic DNA, i.e., containing fragments of DNA isolated from exactly the same mouse strain as that used for obtaining ES cells used.

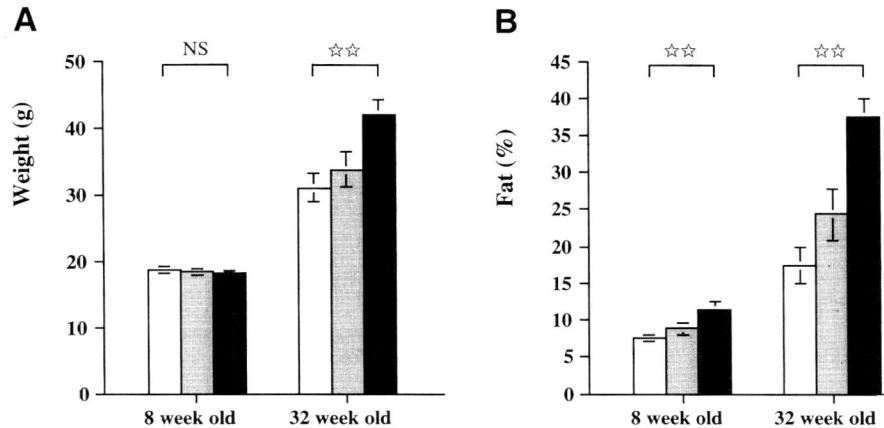

Fig. 2. Body weight **(A)** and body fat **(B)** in female NPY Y1 receptor-deficient mice. Body composition was determined in female mice at 8 and 32 weeks of age. Increased body fat was detected in Y1 knockout mice as early as 8 weeks after birth, although no difference in body weight can be observed at that time. By 32 weeks of age, both body weight and fat are increased in the Y1 knockout group (black). Interestingly, heterozygote animals (gray) are intermediary between wild-type mice (white) and homozygote mutants. This fat accumulation is probably the result of a decreased metabolism in Y1 knockout mice due to lower overall activity in the absence of hyperhpagia *(3)*.

    b.  Isolate homologous DNA fragment of sufficient length (usually about 7–9 kb).
    c.  Replace as much of the coding sequences as possible with a positive select-able marker (e.g., a neomycin resistance cassette).
    d.  If possible, delete all of the coding sequences or at least the 5' exons.
    e.  Introduce a negative selectable marker, such as the thymidine kinase gene, at the end of one arm of the homology (e.g., in **Fig. 1**, it is in the right arm of the targeting vector).

2. This cell culture protocol is best for HM-1 cells. Culture conditions should be adjusted depending on the ES cell line and according to the experience of the laboratory that established it. It is also advisable to test each cell line for sensitiv-ity to selective drugs (e.g., G418).

3. ES cells should be checked for signs of differentiation. If differentiation occurs, cells can be cultured for some passages on a mitomycin C-treated fibroblast monolayer. (Treat almost confluent fibroblasts with 10 µg/mL mitomycin C for 3–5 hours, trypsinize, wash three times, and plate again on gelatinized dishes.)

4. Transfection can be performed using various electroporation settings. Values given here are very gentle to the ES cells. However, one might want to increase the chance of DNA transfer by using more stringent conditions (e.g., 250 V, 250 µF) provided the ES cells can take it.

5. Recent services provided by private companies allow for convenient outsourcing of blastocyst injection.
6. We routinely use a coffee bean grinder for homogenization.

## References

1. Stephens, T. W., Basinski, M., Bristow, P. K., Bue-Valleskey, J. M., Burgett, S. G., Craft, L., et al. (1995) The role of neuropeptide Y in the antiobesity action of the obese gene product. *Nature* **377,** 530–532.
2. Blomqvist, A. G. and Herzog, H. (1997) Y-receptor subtypes: how many more? *Trends Neurosci.* **20,** 294–298.
3. Pedrazzini, T., Seydoux, J., Künstner, P., Aubert, J. F., Grouzmann, E., Beermann, F., and Brunner, H.-R. (1998) Cardiovascular response, feeding behaviour and locomotor activity in mice lacking the NPY Y1 receptor. *Nature Med.* **4,** 722–726.
4. Joyner, A. L. (1993) in *Gene Targeting* (Rickwood, D. and Hames, B. D., eds.), Oxford University Press, Oxford.
5. Entenman, C. (1957) in *Methods in Enzymology* (Colowick, S. P. and Kaplan, O., eds.), Academic, New York, pp. 299–317 .
6. Selfridge, J., Pow, A. M., McWhir, J., Magin, T. M., and Melton, D. W. (1992) Gene targeting using the mouse HPRT minigene/HPRT-deficient embryonic stem cell system: inactivation of the mouse ERCC-1 gene. Somat. *Cell Mol. Genet.* **18,** 325–336.

# III

## ANTISENSE TECHNOLOGY

# 9

## Antisense Oligonucleotide Approach to Study NPY-Mediated Feeding Signal Transduction

### Sulaiman Sheriff, William T. Chance, and Ambikaipakan Balasubramaniam

## 1. Introduction

Antisense oligodeoxynucleotides have been successfully employed to study the function of viral and cellular genes by sequence-specific inhibition. The antisense oligodeoxynucleotides are single-stranded, short, native, or chemically modified nucleotides that inhibit individual gene expression by selective complementary base pairing. These antisense compounds that inhibit gene expression can be grouped into two classes: (1) those leading to a reduction in target RNA levels, and (2) those that do not reduce target RNA levels. Inhibition of RNA function by antisense oligodeoxynucleotides without affecting RNA stability may be due to the disruption of RNA-protein or RNA-RNA interactions that are essential for RNA translation into protein synthesis. Translation arrest without destabilizing mRNA has been demonstrated in vitro using compounds targeted to the translation initiation site *(1,2)*. Binding of antisense oligodeoxynucleotides to the AUG start codon can inhibit translation initiation by inhibiting ribosome assembly *(3)*.

A second mechanism leading to the destabilization of target mRNAs might be the primary pathway of antisense inhibition in in vivo systems. This inhibitory mechanism of antisense oligonucleotides on mRNAs acts through activation of RNAase H, which is an endonuclease that cleaves the RNA strand of RNA-DNA duplexes *(4)*. RNAase H has been shown to degrade the RNA component of oligonucleotide-RNA duplexes in cell-free systems as well as in *Xenopus* oocytes *(5)*. Physiologically, this endogenous endonuclease may act as an antiviral defense for cellular function. The enzyme binds to DNA-RNA

From: *Methods in Molecular Biology, vol. 153: Neuropeptide Y Protocols*
Edited by: A. Balasubramaniam © Humana Press Inc., Totowa, NJ

heteroduplexes in the cytoplasm and degrades the RNA strand, thereby destabilizing the entire mRNA molecule.

Neuropeptide Y (NPY) is a potent stimulator of eating behavior. Our previous studies suggest that NPY-induced food intake may be mediated through hypothalamic NPY receptors coupled to a pertussis toxin-sensitive G-protein *(6)*. All NPY receptors cloned to date have been shown to be coupled positively to intracellular $Ca^{2+}$ and negatively to cyclic adenosine monophosphate (cAMP) pathways *(7)*. Antisense technology may be used to study the role of intracellular signaling cascades. Changes in intracellular $Ca^{2+}$/cAMP lead to the activation of cAMP-responsive element (CRE) binding transcription factors. Upon activation, these transcription factors are translocated to the nuclear compartment, where they bind to CRE, 5'-TGACGTCA-3', present in the regulatory sequence of many eukaryotic genes. This results in the stimulation or repression of the target genes *(2)*. The elements conferring NPY action further downstream are far from clear. Our recent in vivo studies suggest that food deprivation and intrahypothalamic NPY administration induced [$^{32}$P]CRE binding activity and cAMP-responsive element binding protein (CREB) phosphorylation in rat hypothalamus. These intracellular signaling events may be responsible for the central regulation of eating behavior *(8)*. This chapter describes the methodology to identify the CRE-binding transcription factors mediating NPY-induced food intake in rats via an antisense oligodeoxynucleotide approach.

## 2. Materials

1. Stereotaxic instrument (#900, David Kopf Instruments, Tujubga, CA).
2. Cannulae implantation (Plastics One, Roanoke, VA): stainless-steel mounting screws. (0–80 × 3/32, 2.4 mm); quick curing dental acrylic: Hygenic Repair Resin (Hygenic, Akron, OH).
3. Injection (Hamilton, Reno, NV): 50 µL syringe (#705-LT); stainless-steel 31-gage injection needle with Teflon hub (#KF731), repeating dispenser (#PB600-1).
4. Sprague-Dawley rats (Charles River, Willmington, MA).
5. Nitrocellulose membrane 0. 45 µm, 7 × 8.4 cm (Trans-Blot transfer medium, Bio-Rad, Hercules, CA).
6. Nonidet-40 (NP-40; Sigma, St. Louis, MO).
7. Tween-20 (polyoxyethylenesorbitan monolaurate, Sigma).
8. Triton X-100 (t-octylphenoxypolyethoxyethanol, Sigma).
9. Bovine serum albumin (BSA) (Bovine fraction V, RIA grade, Sigma).
10. Acrylamide (Bio-Rad).
11. Bis *N*, *N*'methylene-bis acrylamide (Bio-Rad).
12. Ready gels (10% precast gel for polyacrylamide electrophoresis, 50-µL wells, Bio-Rad).
13. Protein assay dye reagent (Bio-Rad).

14. Medical X-ray film (Fuji).
15. Enhanced chemiluminescence (ECL), Western blotting detection reagents (Amersham, Arlington Heights, IL).
16. CREB antisense, CREB sense phosphorothioate oligodeoxynucleotides, gel filtration grade (The Midland Certified Reagent Company, Midland, TX).
17. Nonfat dry milk (blotting grade, Bio-Rad).
18. CRE containing double-stranded 39-mer oligodeoxynucleotide (Life technologies, Bethesda, MD). This oligo is no longer available from Life technologies; it can be custom synthesized and purified by the Midland Certified Reagent Company.
19. T4 polynucleotide kinase (New England Biolabs, Beverly, MA).
20. $[\gamma^{32}P]ATP$ 6000 Ci/mmol, (NEN, Boston, MA).
21. Salmon sperm DNA (Life Technologies).
22. Sephadex G-50 spin column Pharmacia, Gaithersburg, MD.
23. CREB antibody (raised in rabbit, primary antibody, Santa Cruz Biotechnology, Santa Cruz, CA).
24. Goat anti-rabbit IgG conjugated to peroxidase (Bio-Rad).
25. 3MM blotting paper (Whatman, Hillsboro, OR).
26. Slide-A-Lyzer dialysis cassette (Pierce, Rockford, IL).
27. Phenylmethylsulfonyl fluoride (PMSF; aprotinin, leupeptin, sodium orthovanadate; Sigma).
28. 2-Mercaptoethanol (98%; Sigma).
29. Artificial cerebrospinal fluid (CSF): 129 m$M$ NaCl, 1.3 m$M$ CaCl$_2$, 0.8 m$M$ MgCl$_2$, 2.5 m$M$ NaHCO$_3$, 0.5 m$M$ Na$_2$HPO$_4$, 3.0 m$M$ KCl. The pH of the solution was adjusted to 7.4, filtered on 0.22-mm nitrocellulose filter and stored at 4°C.
30. Homogenization buffer: 20 m$M$ HEPES, pH 7.4, 50 m$M$ β-glycerophosphate, 2 m$M$ EGTA, 1 m$M$ sodium vanadate, 1 m$M$ PMSF, and 2 µg/mL aprotinin.
31. Lysis Buffer: 20 m$M$ HEPES, pH 7.9, 400 m$M$ NaCl, 5 m$M$ MgCl$_2$, 0.1 m$M$ EGTA, 5 m$M$ dithiothreitol (DTT), 1 m$M$ PMSF, 1 m$M$ sodium vanadate, 10 m$M$ sodium pyrophosphate, 1% NP-40, 20% glycerol, 10 µg/mL leupeptin, and 1 µg/mL pepstatin.
32. TE buffer: 10 m$M$ Tris, pH 7.5, 1 m$M$ EDTA.
33. 0.25X TBE buffer: 50 mL 5X TBE in 1 L H$_2$O. (5X TBE buffer = 500 m$M$ Tris-HCl, pH 8.0, 450 m$M$ borate, 5 m$M$ EDTA).
34. Tris-glycine sodium dodecyl sulfate (SDS) buffer: 25 m$M$ Tris-HCl , 0.192 $M$ Glycine, 0.1% SDS (w/v), pH 8.3.
35. 2X Laemmli buffer: 100 m$M$ Tris-HCl, pH 6.8, SDS (4%), 20% glycerol, 0.01% bromophenol blue, 200 m$M$ DTT or 1.25 $M$ 2-mercaptoethanol. This buffer lacking DTT or 2-mercaptoethanol can be stored at room temperature (RT). DTT or 2-mercaptoethanol should be added just before use from 1 $M$ (DTT) or 12.54 $M$ (2-mercaptoethanol) stocks.
36. Tris-methanol transfer buffer: 25 m$M$ Tris base, 0.192 $M$ glycine, 20% methanol (v/v), pH 8.5. Methanol should be added at the time of use.
37. Tris-buffered saline (TBS): 20 m$M$ Tris base, 137 m$M$ NaCl, adjust pH to 7.6 using. 1 $M$ HCl (approx 3.8 mL/L).

38. Tris-buffered Tween (TBS-T): to prepare 1 L of TBS-T, add 1 mL of Tween (100%) to 1 L of TBS buffer.
39. Blocking buffer: to prepare 100 mL of blocking buffer, add 5 g nonfat dry milk to 100 mL of TBST and mix well in a rotary mixer. This buffer needs to be prepared fresh.
40. Primary antibody (CREB) dilution buffer: to prepare 20 mL of antibody dilution buffer, add 1 g of BSA to 20 mL of TBST and mix well . This buffer needs to be prepared fresh.
41. Buffer A: 20 m*M* HEPES, pH 7.9, 25% glycerol, 1.5 m*M* $MgCl_2$, 20 m*M* KCl, 0.2 m*M* EDTA, 1 m*M* DTT, 1 m*M* vanadate, 1 m*M* PMSF, 2 μg/mL aprotinin.
42. Buffer B: buffer A containing 0.8 *M* KCl.
43. Dialysis buffer: 20 m*M* HEPES, pH 7.9, 20% glycerol, 100 m*M* KCl, 0.2 m*M* EDTA, 1 m*M* DTT, 1 m*M* vanadate, 1 m*M* PMSF, 2 μg/mL aprotinin.
44. Sample buffer (5X) for gel-shift assay: 50 m*M* Tris-HCl, pH 7. 5, 500 m*M* NaCl, 5 m*M* DTT, 5 m*M* EDTA, 20% glycerol (v/v), and 0.4 mg/mL sonicated salmon sperm DNA. Aliquots of this buffer can be stored at –20°C for at least 3 months.

## 3. Methods

### 3.1. Surgery and Implantation of Perifornical Hypothalamic Cannula

House adult male Sprague-Dawley rats weighing 250–350 g, are kept individually in plastic shoe box cages in a temperature- and humidity-controlled environment under a 12-h light/dark cycle with free access to Purina Rat Chow and water. Acclimatize rats to laboratory conditions for 1 week before the onset of experiments. Anesthetize rats using ketamine/xylazine (60–80 mg/kg, im), and implant 24-gage stainless steel cannulae into the perifornical hypothalamus (PFH) using the following stereotaxic coordinates taken from the interaural line: anterior +6.0 mm, lateral 1.0 mm, ventral +2.5 mm (upper incision bar = +2.7 mm). Allow 2-week recovery period before initiating the in vivo studies.

### 3.2. Administration of Antisense Oligodeoxynucleotides

Use phosphorothioate-modified oligo directed at the translation start site of the CREB mRNA as reported previously *(9)* (*see* **Note 1**). Inject 15 μg of gel filtration grade CREB antisense oligodeoxynucleotide 5'-TGGTCATCTAGTCACCGGTG-3' in 2 μL of artificial CSF into the PFH of seven rats. Similarly, inject 15 μg of CREB sense oligodeoxynucleotide 5'-CACCGGTGACTAGATGACCA-3' in 2 μL of artificial CSF to another group of seven rats. Inject 2 μL of CSF to control rats. After the injections, test the effect of 1 μg NPY on 1 h food intake on the following day and weekly for next 2 weeks, (*see* **Note 2**). Seven days following the last NPY test, inject 15 μg of CREB sense or antisense oligos into PFH of the same group of rats (*see*

**Note 3**). On the following day, inject 1 µg of NPY the and sacrifice the rats 60 min later by decapitation.

### 3.3. Isolation of Hypothalamus

Isolate the whole hypothalamus from the ventral surface of the brain by making an incision 2.5 mm deep from just anterior to the optic chiasm and extending to the posterior mammillary area and laterally to the choroid fissure. Freeze the hypothalamus in liquid nitrogen, and store at –80°C.

### 3.4. CRE Gel-Shift Assay
### or CRE Binding Protein Detection Assay

### 3.4.1. Nuclear Protein Extraction

Prepare nuclear protein extracts of the hypothalamus as described by Sheriff et al. *(8)*. All procedures should be carried out at 4°C.

1. Transfer two hypothalami from each treatment to 12 × 75-mm glass tubes (pre-cooled in dry ice).
2. Add 150 µL of homogenization buffer in each tube.
3. Homogenize the hypothalami in Polytron homogenizer for 15 s with a speed setting at 3.
4. Keep the resulting homogenate in ice for 15 min.
5. Centrifuge the homogenate at 3300*g* for 15 min at 4°C.
6. Discard the supernatant, add 0.5 volume of buffer A to 1 volume of the pellet, and rehomogenize as **step 3**.
7. Add 0.5 volume of buffer B to the homogenate.
8. Mix the homogenate continuously in a rotary mixer for 1 h.
9. Carefully transfer the homogenate to a Slide-A-Lyzer dialysis cassette with 10,000 $M_r$ cutoff using a 1-mL syringe with a 21-gage needle.
10. Dialyze the Slide-A-Lyzer dialysis cassette containing sample in 500 volumes of dialysis buffer for 8 h.
11. Centrifuge the suspension at 13,000*g* for 30 min.
12. Save the supernatant, and estimate its protein content using the Bio-Rad protein assay kit.
13. Adjust the protein concentration of the samples to 5 µg/µL.
14. Freeze the samples on dry ice, and store at –80°C.

### 3.4.2. Preparation of [32]P-Labeled Oligonucleotide Probe

We have been using double stranded 39-mer oligonucleotide containing a tandem repeat of the consensus sequence for the CRE containing DNA binding site to increase the sensitivity of the DNA protein interaction. Use the sequence of the synthetic 39-mer double-stranded oligo described below at a concentration of 5 mg/mL in TE buffer.

**5'**-<u>GATC</u>**TGACGTCA**<u>TGAC</u>**TGACGTCA**<u>TGAC</u>**TGACGTCA**<u>TCA</u>-**3'**

**3'**-**ACTGCAGT**<u>ACTG</u>**ACTGCAGT**<u>AC</u> **TG**<u></u>**ACTGCAGT**<u>AGTCTAG</u>-**5'**

1. Add 1 μL (5 ng) of double-stranded oligo, 2.5 μL 10X polynucleotide kinase buffer, 10 μL fresh lot of [γ$^{32}$P]ATP (6000 Ci/mmol), 10.5 mL sterile H$_2$O, and 1 μL T4 polynucleotide kinase in a 1.5-μL microcentrifuge tube.
2. Incubate this mixture for 45 min at 37°C.
3. Add 50 μL of 50 m*M* Tris-HCl, pH 7.5, buffer.
4. Carefully transfer the entire material on an equilibrated sephadex G-50 spin column.
5. Centrifuge the column at 1500*g* for 3 min at RT.
6. Collect eluant in a clean 1.5-mL microcentrifuge tube.
7. Determine the efficiency of labeling by counting a 1-μL aliquot in a scintillation counter. (approx 1-μL will yield 150,000–300,000 cpm).

### 3.4.3. Preparation of Gel (6% Nondenaturing Polyacrylamide Gel)

1. Make a vertical glass sandwich using 11 × 16-cm clean glass plates with a 0.75-mm-thick spacer using clamps.
2. Lay down the glass plate sandwich horizontally on a clean even surface.
3. Make 30 mL of 6% nondenaturing polyacrylamide gel by mixing 6 mL of polyacrylamide/bis acrylamide (30:0.8% w/v), 22.1 mL H$_2$O, 0.9 mL glycerol, 1.5 mL 5X TBE, 15 μL *N,N,N,N*-trimethyle-nediamine (TEMED), 0.3 mL of 10% (W/V) ammonium persulfate (freshly made or frozen aliquot) in a clean 100-mL glass flask.
4. Mix well using 10-mL pipet, and remove air bubbles, if any.
5. Add the liquid gel mixture as quickly as possible using a 25 mL pipet by holding the glass sandwich at a 45° angle.
6. Fill the entire glass plate with the gel mixture with no air bubbles.
7. Insert a 0.8 mm thick 10–12-well comb from the top into the gel mixture.
8. Repeat **step 2**.
9. Cover the top surface of the glass sandwich with clean Saran Wrap.
10. Allow the gel to polymerize at RT for 30–45 min.

### 3.4.4. Sample Preparation for Gel-Shift Assay

1. Thaw sample protein from –80°C on ice.
2. Mix 2 μL sample protein (10 μg), 2 μL 5X sample buffer, and 5.5 μL H$_2$O in a clean microcentrifuge tube (*see* **Note 4**).
3. Incubate this mixture at 4°C for 15 min.
4. Add 0. 5 μL $^{32}$P-labeled probe (approx 75,000–100,000 cpm).
5. Incubate at RT for 20 min. This tube will represent the binding assay tube.
6. In a parallel tube, add 1 μL of 50–100-fold excess unlabeled oligonucleotide (5 ng) as a competitor, and mix all the other reagents as in **step 2** (2 μL sample protein, 2 μL 5X sample buffer, 4.5 mL H$_2$O). This tube will represent the competition assay tube.
7. Repeat **steps 3–5**.

8. Add 0.5 mL of 0.1% bromophenol blue dye.
9. Fix the glass plates containing polymerized gel to the vertical gel apparatus.
10. Add 750–800 mL of 0.25X TBE running buffer into the tank.
11. Flush the wells with the running buffer, and remove the air bubbles at the area of gel contacts to the running buffer in the lower tank.
12. Prerun the gel at 4°C for 15–30 min.
13. Load the samples into wells.
14. Run the gel at 4°C at 150 V until dye front is 2 cm from the bottom of the gel (3.5–4 h).
15. Carefully transfer the gel to Whatman 3MM blotter paper, cover with Saran Wrap, and dry under vacuum at 80°C for 1 h.
16. Expose the gel to an X-ray film using an intensifying screen at –70°C overnight.

### 3.4.5. Running the Protein Gel

1. Homogenize the treated hypothalami individually in 100 µL lysis buffer in 12 × 75-mm glass tubes using a Polytron homogenizer at 4°C for 15 s with the speed setting at 3.
2. Transfer the homogenate to a 1.5-mL microcentrifuge tube.
3. Mix the homogenate continuously in a rotor at 4°C for 30 min.
4. Centrifuge the homogenate at 15,000$g$ for 30 min at 4°C.
5. Transfer the supernatant to a new tube, estimate the protein content using Bio-Rad's reagent, and adjust it to 4 µg/mL.
6. Mix the supernatant with an equal volume of 2X Laemmli buffer.
7. Heat the samples at 95°C for 3–5 min.
8. Load 50 µL of diluted samples (100 µg protein) into the wells and 10 µL of pre-stained broad-range molecular weight markers in a separate lane of 10% minigel.
9. Run samples on gel for 1 h at 200 V.
10. Stop the run once the bromophenol dye has reached the bottom of the gel.

### 3.4.6. Semi-Dry Transfer of Proteins to Nitrocellulose Membrane

1. Soak the blotter paper (Whatman 3MM) and nitrocellulose membrane in Tris-methanol transfer buffer for 3–5 min (until they are saturated).
2. Place two saturated blotter papers on the (Bio-Rad) transfer apparatus.
3. Make a gel and saturated nitrocellulose membrane sandwich and place it on top of the blotter paper by keeping the gel up and membrane down.
4. Place two saturated blotter papers on top of the gel.
5. Keep the lid on the unit, and connect the leads to the power supply.
6. Run the transfer for 1 h at 25 V (*see* **Note 5**).
7. Determine the electrotransfer of proteins by the transfer of prestained molecular weight markers to nitrocellulose membrane.

### 3.4.7. Immunoblot (Western Blot) Analysis

After the protein transfer to nitrocellulose membrane:

1. Rinse the membrane gently in TBS buffer on a platform shaker for 3 min at RT.
2. Incubate the membrane in blocking buffer for 1 h at RT.

3. Wash the membrane three times for 5 min each with TBST.
4. Transfer the membrane to CREB antibody solution, diluted (1:1000) in primary antibody dilution buffer, and incubate at 4°C for 16 h with gentle agitation on a platform shaker (*see* **Note 5**).
5. Wash the membrane three times in blocking buffer for 15 min each at RT.
6. Incubate the membrane with second antibody (horseradish peroxidase-conjugated goat anti-rabbit IgG) diluted at 1:3000 in blocking buffer for 1 h with gentle agitation at RT.
7. Repeat **step 5**.
8. Wash the membrane once in TBS for 5 min at RT.
9. Mix an equal volume of detection solution 1 (5 mL) with detection solution 2 (5 mL).
10. Drain the membrane from the TBS, and transfer it to a piece of Saran Wrap by facing the protein side up using a clean blunt forceps. (Do not allow the membrane to dry.)
11. Add the mixture solution (from **step 9**) on top of the membrane, and incubate for exactly 1 min at RT without agitation.
12. Drain off the mixture solution completely from the membrane, and wrap it in a piece of Saran Wrap with the protein side facing up.
13. Transfer the membrane to a film cassette as quickly as possible.
14. Place an autoradiography film on top of the membrane in a dark room for 30 s.
15. Remove the film and develop it.
16. Immediately replace with a fresh piece of unexposed film, reclose the cassette for 1 min, and develop the film.
17. Repeat **step 16** for 3 min with unexposed new film, and develop it.
18. A protein band will appear with a molecular weight of 43 kDa.
19. Select the best film from the three, based on optimal signal intensity and minimal background.

## 4. Notes

1. Phosphorothioate oligodeoxynucleotides of gel filtration grade worked well in our hands. Many investigators use antisense oligodeoxynucleotides from 16 to 24 nucleotides in length. A minimum of 16 nucleotides is essential to exhibit unique binding properties within the genome. An increase in length of the oligo may help in stable heteroduplex formation; however, this may decrease specificity by forming more mismatched double-stranded complexes and may also decrease penetration into the cell *(2)*.
2. CREB antisense treatment had an immediate detrimental effect on ad libitum food intake. However, 6 days after treatment, CREB antisense-treated rats returned to normal eating behavior in comparison with sense- or artificial CSF-treated rats. The reduction in food intake was reflected in an immediate loss of body weight, which plateaued with the return of normal feeding (**Fig. 1**). These experiments suggest that endogenous CREB protein may regulate the feeding behavior and that the inhibitory effect of CREB on food intake is reversible.

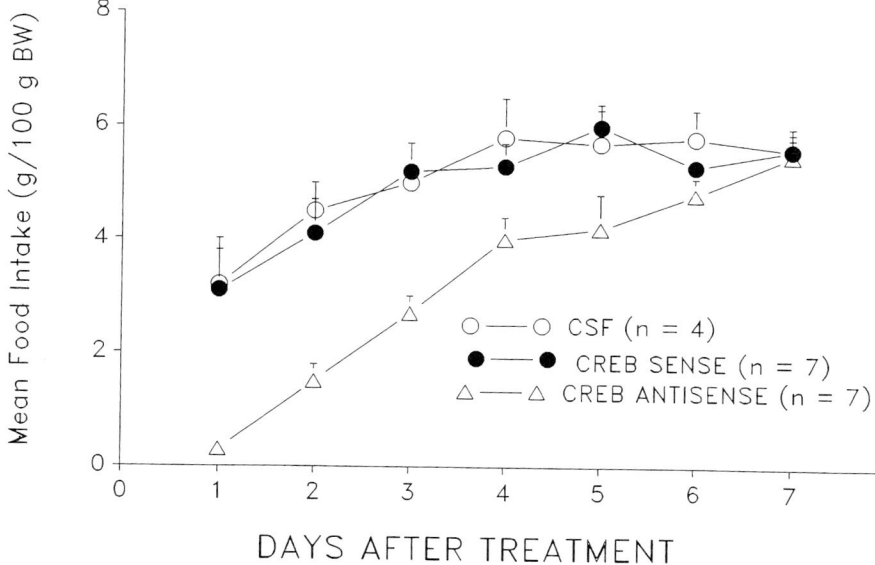

Fig. 1. Mean (± SEM) daily intake of rat chow following intrahypothalamic injection of artificial CSF (2 μL), CREB sense oligonucleotides (15 μg in 2 μL CSF), or CREB antisense oligonucleotide (15 μg in 2 μL CSF).

3. NPY-induced feeding had different results, depending on the temporal distance from antisense treatment. Thus, although NPY elicited significant feeding 24 h after treatment with CREB sense oligodeoxynucleotide, this food intake was not different from that following injection of CSF, suggesting the downstream role of CREB in mediating NPY action. A similar reduction of NPY-induced feeding was observed 7 days later, when ad libitum food intake had normalized. Two weeks after injection of CREB antisense, however, normal NPY feeding had returned to treated rats (**Fig. 2**).

4. Repeated freezing and thawing or improper storage of 5X sample buffer may allow degradation of salmon sperm DNA and cause nonspecific binding with nuclear proteins. This may result in increased retention of sample in the well. Two species of protein/[$^{32}$P]CRE complexes well observed on the gel. The intense, slowly migrating protein DNA complex may represent the dimers, while fast moving species may represent the monomeric forms of CRE binding proteins. The free and unbound [$^{32}$P]CRE oligo was observed at the bottom of the gel. Gel-shift assay for CRE binding showed a reduction in [$^{32}$P]CRE binding activity in the hypothalamic nuclear proteins from the CREB antisense/NPY-treated group in comparison with the CREB sense/NPY or CSF/NPY groups (**Fig. 3**). Densitometric analysis of gel-shift X-ray images revealed 164 arbitrary densitometric units (ADU) in the antisense-treated rats in comparison with CSF/

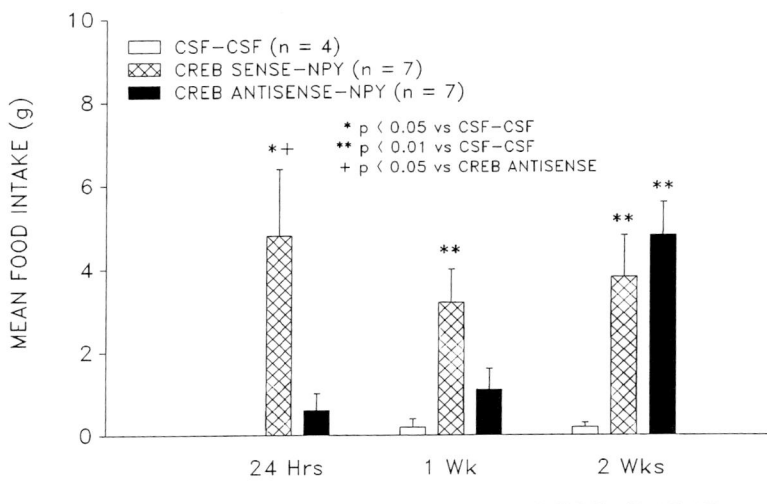

Fig. 2. Mean (± SEM) 1-h intake of rat chow following injection of artificial CSF (1 μL) or NPY (1 μg) into the hypothalamus of rats treated previously, either for 24 h, 1 week, or 2 weeks with CSF (2 μL), CREB sense oligodeoxynucleotide (15 μg in 2 μL CSF) or CREB antisense oligonucleotide (15 μg in 2 μL CSF).

Fig. 3. CRE DNA binding activity to hypothalamic nuclear proteins obtained from rats 60 min after intrahypothalamic injection of artificial CSF (1 μL) or NPY (1 μg). The rats were pretreated (24 h) with artificial CSF (2 μL), CREB sense oligodeoxynucleotides (15 μg in 2 μL CSF), or CREB antisense oligodeoxynucleotides (15 μg in 2 μL CSF).

NPY-treated rats (328 ADU). Rats from the CREB sense oligo-treated group also exhibited a weaker, 18% decrease (269 ADU) in [$^{32}$P]CRE binding activity. No DNA binding activity was observed in the probe lane, as no protein was loaded. The specificity of CRE binding was verified by the complete inhibition of

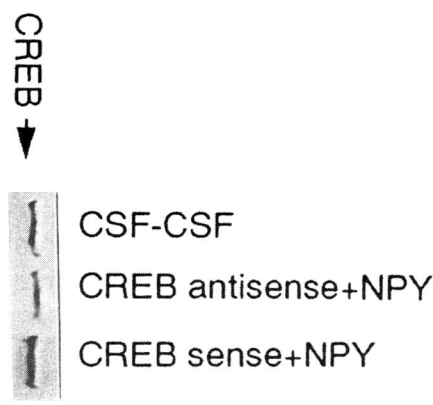

Fig. 4. CREB protein immunoreactivity in rat hypothalamus euthanized 60 min after injection of artificial CSF (1 μL) or NPY (1 μg in 1 μL CSF). The rats were pretreated (24 h) with CSF (2 μL), CREB sense oligodeoxynucleotides (15 μg in 2 μL CSF), or CREB antisense oligodeoxynucleotides.

$[^{32}P]$CRE/protein complex by preincubating 100-fold excess unlabeled CRE oligo in the binding reaction (lane 5).

5. Nitrocellulose membranes are better than PVDF membrane for protein transfer. Transferring the proteins to membranes for 45 min to 1 hour is ideal; longer transfer time may distort gel bands or denature proteins. Using the CREB antibody in 5% BSA containing TBST buffer greatly minimizes the background signal. A protein with a molecular mass of approx 43 kDa was observed on the film, which represents the CREB protein. Determination of CREB protein level in the hypothalamus showed elevated CREB in the sense oligonucleotide treatment following injection of NPY (78 ADU), compared with CREB antisense/NPY-treated rats (43 ADU). No significant change was observed in the CSF/CSF-treated rats (41 ADU) (**Fig. 4**).

## Acknowledgments

This work was supported in part by the Public Health Service Grants DK-53548 (to S. S.) and GM47122 (to A. B.) and by a grant from the Department of Veterans Affairs (to W. T. C.).

## References

1. Monia, B. P., Lesnik, E. A., Gonzalez, C., Lima, W. F., McGee, D., Guinosso, C. J., et al. (1993) Evaluation of 2'-modified oligonucleotides containing 2'-deoxy gaps as antisense inhibitors of gene expression. *J. Biol. Chem.* **268,** 14,514–14,522.
2. Russell, D. S., Widnell, K. L., and Nestler, E. J. (1996) Antisense oligonucleotides: new tools for the study of brain function. *Neuroscientist* **2,** 79–82.

3. Shakin, S. H. and Liebhaber, S. A. (1986) Destabilization of messenger RNA/complementary DNA duplexes by elongating 80 S ribosome. *J. Biol. Chem.* **261,** 16,018–16,025.
4. Crouch, R. J. and Dirksen M. L (1982) In Nucleases (Linn, S. M. and Roberts, R. J., eds.), Cold Spring Harbor Laboratory, Cold Spring Harbor, NY, pp. 211–241.
5. Doris-Keller, H. (1979) Site-specific enzymatic cleavage of RNA. *Nucleic Acids Res.* **7,** 179–192.
6. Chance, W. T. S., Sheriff, S., Foley-Nelson, T., Fischer, J. E., and Balasubramaniam, A. (1989) Pertussis toxin inhibits neuropeptide Y induced feeding in rats. *Peptides* **10,** 1283–1286.
7. Balasubramaniam, A. (1996) Neuropeptide Y family of hormones: receptor subtypes and antagonists. *Peptides* **18,** 445–457.
8. Sheriff, S., Chance, W. T., Fischer, J. E., and Balasubramaniam, A. (1997) Neuropeptide Y treatment and food deprivation increase cyclic AMP response element-binding in rat hypothalamus. *Mol. Pharmacol.* **51,** 597–604.
9. Konardi, C., Cole, R. L., Heckers, S., and Hyman S. E. (1994) Amphetamine regulates gene expression in rat striatum via transcription factor CREB. *J. Neurosci.* **14,** 5623–5634.

# 10

## A Gαi RNA-Antisense Expression Strategy to Investigate Coupling of Peptide YY/Neuropeptide Y Receptor to Gαi

### Thierry Voisin, Anne-Marie Lorinet, Mathieu Goumain, and Marc Laburthe

### 1. Introduction

Following its discovery in rat intestinal epithelial cells *(1)*, the peptide YY (PYY) receptor has been characterized in the proximal tubule PKSV-PCT cell line derived from kidneys of SV40 large T antigen transgenic mice *(2)*. This receptor is PYY preferring since it binds the intestinal hormone PYY *(3)* with high affinity and neuropeptide Y (NPY) *(3)* with a 5–10-fold lower affinity *(1–3)*. PYY and NPY trigger several biological effects through interaction with PYY receptors including inhibition of adenylyl cyclase activity *(4)*, and Cl⁻ secretion in small intestine *(3)*, inhibition of lipolysis in fat cells *(5)*, and stimulation of epithelial cell growth *(2,6)*. PYY receptor pharmacology is clearly different from that of other receptors for the PP-fold family of peptides, including the Y1, Y3, Y4, Y5, and Y6 subtypes. The PYY receptor resembles the Y2 subtype of NPY receptor, which does not discriminate between PYY and NPY, but, like the PYY receptor, binds with high affinity long COOH-terminal fragments of PYY or NPY *(4,7)*. The intestinal PYY receptor has not been cloned yet, but it has been characterized as an $M_r$ 44,000 glycoprotein by crosslinking experiments and hydrodynamic studies *(8)*.

PYY receptors are characterized in the PKSV-PCT cell line *(2)* derived from microdissected proximal convoluted tubules of kidneys from transgenic mice harboring the SV40 large T antigen placed under the control of the rat L-type pyruvate kinase 5'-regulatory sequence *(9,10)*. PYY receptor-mediated events are triggered through interaction of PYY receptors with pertussis toxin-sensitive Gi-proteins in PKSV-PCT cells. Indeed, preincubation of cells with per-

From: *Methods in Molecular Biology, vol. 153: Neuropeptide Y Protocols*
Edited by: A. Balasubramaniam © Humana Press Inc., Totowa, NJ

tussis toxin completely reverses the PYY-induced inhibition of cyclic adenosine monophosphate (cAMP) production and stimulation of cell growth and converts PYY receptors to a low-affinity state *(2)*. In view of the facts that (1) multiple Gi-proteins including Gi1, Gi2, and Gi3 can contribute to heptahelical receptor-mediated inhibition of adenylyl cyclase *(11)*; and (2) pertussis toxin-sensitive G-proteins have been shown to be crucial for the mitogenic action of several agents *(12)*, characterization of the pertussis toxin-sensitive Gi-protein(s) coupled to PYY receptors is an important step that will lead to further understanding of the mechanism of action of PYY.

We have developed stable expression of antisense $G\alpha i$ RNA in a renal proximal tubule PKSV-PCT Cl.10 clone isolated from the parent PKSV-PCT cells in order to identify Gi-proteins coupled to PYY receptors. PKSV-PCT Cl.10 cell clones were obtained in which endogenous $G\alpha i$ proteins are permanently downregulated after transfection with antisense $G\alpha i$ expression vectors. By studying receptor affinity, inhibition of cAMP production, and stimulation of cell growth, we provide evidence for the exclusive coupling of the PYY receptor to the Gi2-protein.

## 2. Materials

### 2.1. Cultured Cells

1. Dulbecco's modified Eagle's medium (DMEM)/Ham's F-12 1:1 (v/v; Gibco-BRL, Cergy Pontoise, France) stored at 4°C.
2. Fetal calf serum (Gibco-BRL) stored at –20°C.
3. Standard culture medium, pH 7.4, stored at 4°C: DMEM/Ham's F-12 1:1 (v/v), 2% fetal calf serum (v/v) with 60 n$M$ sodium selenate, 5 µg/mL transferrin, 2 m$M$ glutamine, 50 n$M$ dexamethasone, 1 n$M$ triiodothyronine, 10 n$M$ epidermal growth factor, 20 m$M$ HEPES, 5 µg/mL insulin, and 20 m$M$ D-glucose (Sigma).
4. 10X phosphate-buffered saline (PBS) without $Ca^{2+}$ and $Mg^{2+}$ (PBS, 1.3 $M$) (Gibco-BRL) stored at 4°C.
5. Versene 1:5000 with 10% trypsine (Gibco-BRL) stored at 4°C.
6. Pertussis toxin (Sigma) stored at 4°C.

### 2.2. Antisense $G\alpha i$ Subunit Expression Vector Constructions

1. pcDNA3/neo expression vector (Invitrogen, Groningen, The Netherlands) stored at –20°C.
2. The 39 bases of the 5'-noncoding region, immediately upstream of end including the ATG translation initiation codon of $G\alpha i2$ (5'-GCGTGTGGGGGCCA-GGCCGGGCCGGCGGACGGCAGGATG-3') and $G\alpha i3$ (5'-GCGAGCCAG-GGCCCGGTCCCCTC TCCGGCCGCCGTCATG-3') and their complementary strand are synthesized commercially (Gibco-BRL).
3. *Eco*RV restriction endonuclease (10 U/µL) (Gibco-BRL) stored at –20°C.

4. 10X *Eco*RV reaction buffer: 500 m$M$ Tris-HCl, 100 m$M$ MgCl$_2$, and 500 m$M$ NaCl, pH 8.0, (Gibco-BRL, Cergy Pontoise, France), stored at –20°C.
5. T4 DNA ligase (1 U/µL).
6. 10X ligation buffer: 300 m$M$ Tris-HCl, pH 7.8, 100 m$M$ MgCl$_2$, 100 m$M$ dithiothreitol (DTT), 10 m$M$ ATP (Promega, Lyon, France) stored at –20°C.
7. Alkaline phosphatase (1 U/µL).
8. 10X dephosphorylation buffer: 500 m$M$ Tris-HCl, 1 m$M$ EDTA, pH 8.5 (Boehringer Mannheim, Meylan, France) stored at 4°C.
9. T4 polynucleotide kinase: 5–10 U/µL
10. 10X kinase buffer: 700 m$M$ Tris-HCl, pH 7.6, 100 m$M$ MgCl$_2$, 50 m$M$ DTT (Promega) stored at –20°C.
11. *Escherichia coli* XL-1 competent cells (Statagene, La Jolla, CA) stored at –80°C.
12. Ethidium bromide 0.625 mg/mL (Interchim, Montluçon, France), stored at room temperature (RT).
13. 0.5X Tris-boric acid-EDTA (TBE) (Interchim, Montluçon, France) stored at RT.
14. 50 mg/mL ampicillin hydrochloride in H$_2$O (Gibco-BRL, Cergy Pontoise, France), stored at –20°C.
15. Luria-Bertani (LB)-agar medium: 10 g/L tryptone, 5 g/L yeast extract, 10 g/L NaCl, and 15 g/L agar.
16. LB- medium: 10 g/L tryptone, 5 g/L yeast extract, and 10 g/L NaCl (Bio 101, Vista, CA) stored at RT.
17. Nucleobond AX 100 Kit (Macherey-Nagel, Hoerdt, France).
18. Sequenase Version 2.0 DNA Sequencing Kit (Amersham Life Science, Les Ulis, France).

## 2.3. Transfection of Antisense Gαi Subunit Expression Vector

1. Gene pulser Electroporator II (Invitrogen, Carlsbad, CA).
2. Electroporation cuvettes 0.4 cm (Invitrogen).
3. Penicillin–streptomycin (10,000 IU/mL–10,000 µG/mL) (Gibco-BRL) stored at –20°C.
4. Geneticin (G418) (Gibco-BRL, Cergy Pontoise, France) stored at –20°C.

## 2.4. Preparation of Particulate Fraction of Cultured cells

1. 1 $M$ HEPES buffer, pH 7.4 (Gibco-BRL) stored at –20°C.
2. 10X PBS⁻ (1.3 $M$; Gibco-BRL) stored at 4°C.
3. A rubber policeman.

## 2.5. Immunoblotting of Gαi subunit of Gi-proteins

1. Acrylamide 4X, bis-acrylamide 2X (37.5:1 mixture) 40% (w/v; Quantum) stored at 4°C.
2. Sodium dodecyl sulfate (SDS) 20% (Quantum) stored at 4°C.
3. Ammonium persulfate (Quantum) stored at 4°C.
4. *N,N,N',N'*-tetramethyl-ethylenediamine (TEMED) (Sigma) stored at 4°C.
5. Electrophoresis sample buffer: 0.06 $M$ Tris-HCl, pH 6.8, 10% gycerol, 3% SDS, and 0.1% bromophenol blue stored at RT.

6. Running buffer: Tris base 0.05 M, glycine 0.4 *M*, and SDS 0.1% stored at RT.
7. Molecular weight marker proteins: myosin (200,000), phosphorylase b (97,000), bovine serum albumin (68,000), ovalbumin (43,000), and carbonic anhydrase (29,000) (MW-SDS-200; Sigma) stored at –20°C.
8. Anti-G$\alpha$i2/G$\alpha$i1 (AS7, dilution 1:1,000) and anti-G$\alpha$i3/G$\alpha$o (EC2, dilution 1:1,000) from NEN (Paris, France), stored at –80°C.
9. Anti-rabbit Ig from donkey, $^{125}$I-labeled, F(ab')$_2$ fragment (IM1340; Amersham) stored at 4°C.
10. Transfer buffer: dissolve 5.82 g Tris, 2.93 g glycine, and 3.75 mL 10% SDS in water, add 200 mL of methanol, and adjust volume to 1 L with water. Store at RT.
11. Trans-Blot Semi-Dry electrophoretic transfer cell (170-3940; Bio-Rad, Ivry sur Seine, France) and cellulose nitrate membrane BA 85 0.45 µm (Schleicher & Schuell, Dassel, Germany).
12. Antibody buffer A: antibody buffer B (50 m*M* Tris-HCl, pH 8.0, 2 m*M* CaCl$_2$, 80 m*M* NaCl, 5% [w/v] nonfat dry milk, 0.02% NaN$_3$) containing 0.2% NP-40 (Sigma), store at 4°C.

### 2.6. PYY Binding to Membrane Receptors

1. Binding incubation buffer: 20 m*M* HEPES, pH 7.4, 2% (w/v) bovine serum albumin (BSA), 17 mg/L phenylmethylsulfonyl fluoride (PMSF), 10 mg/L TLCK, 10 mg/L pepstatin, 10 mg/L leupeptin and 100 mg/L bacitracin, stored at –20°C.
2. Washing buffer: 20 m*M* HEPES, pH 7.4, with 10% sucrose stored at –20°C.
3. Porcine PYY (Neosystem, Strasbourg, France) stored at –20°C.
4. [$^{125}$I]Na (IMS300; Amersham), chloramine T, and sodium disulfite Na$_2$S$_2$O$_5$ (Sigma) stored at RT.
5. Bio-Rad protein assay kit stored at 4°C.

### 2.7. Cyclic AMP Measurement

1. Forskolin (Sigma) stored at –20°C.
2. 3-Isobutyl-1-methylxanthine (IBMX) (Sigma) stored at –20°C.
3. 95% ethanol/5% formic acid precipitation solution stored at 4°C.
4. Radioimmunoassay Pasteur kit (Laboratoires Pasteur, Paris, France).

### 2.8. [Methyl-$^3$H] Thymidine Incorporation

1. EDTA 2.7 m*M* (Gibco-BRL) stored at –20°C.
2. [Methyl-$^3$H]thymidine (Amersham) stored at 4°C.
3. Trichloroacetic acid 5% (Sigma) stored at RT.

## 3. Methods

### 3.1. Cultured Cells

The PKSV-PCT cell line is derived from microdissected proximal convoluted tubules from the kidney of a transgenic mouse (L-PK/Tag1) carrying the

SV40 large T and small t antigens placed under control of the rat L-type pyruvate kinase promoter *(9,10)*. PKSV-PCT cells are cloned by limiting dilution. Briefly, a monodispersed cell suspension is distributed to microtest plates at a mean ratio of 0.25 cells/well. Those wells containing only one cell, as ascertained by microscopic inspection by two independent observers, are identified with their coordinates on the plates. Cells grown in wells observed to contain initially one cell are subsequently transferred to increasingly larger culture vessels. Among the 12 clones obtained, the clone Cl.10 is selected on the basis of its high binding capacity for PYY.

1. Cl.10 cells are cultured in a standard culture medium at 37°C in 5% $CO_2$–95% air atmosphere. As previously shown for the parent cell line PKSV-PCT (2, 9, 10), such culture conditions with D-glucose-enriched medium favors the activation of T antigen transcripts and cell growth. All studies on the Cl.10 cell line are performed between the 4th and 12th passages on sets of cells seeded on plastic culture flasks (25 cm² or 75 cm² surface).

2. The cells are routinely passaged every 7 days:
   a. Wash Cl.10 cells seeded on 25 cm² plastic flasks with 5 mL of PBS⁻ (0.13 *M*) buffer.
   b. Remove the washed cells by incubation at 37°C with 1 mL of Versene 1:5000 with 10% trypsine.
   c. Centrifuge for 5 min (2000*g*) at room temperature to collect the cells.
   d. Resuspend cells in 5 mL of fresh culture medium at 37°C.
   e. Seed 400,000 Cl.10 cells on a new 25 cm² plastic flask with 5 mL of a standard culture medium at 37°C in 5% $CO_2$–95% air atmosphere.
   f. Standard culture medium is changed every 2 days.

3. In some sets of experiments, confluent cells grown in 25-cm² plastic flasks are treated overnight (16 h) with 0.4 µg/mL of pertussis toxin, which is added to the culture medium. A particulate fraction is then prepared as described below and used for binding experiments. A similar procedure is applied to sets of confluent cells before cAMP assay.

## 3.2. Antisense Gαi Subunit Expression Vector Constructions

The pcDNA3 expression vector is used to construct antisense Gαi subunit expression vectors. It contains enhancer/promoter sequences of the human cytomegalovirus intermediate early gene and a polyadenylation signal from the bovine growth hormone gene, an ampicillin resistance gene, and a Col E1 origin of replication for selection and maintenance in *E. coli*, a neomycin-resistant gene expressed from the SV40 early promoter for selection of stable transformants in the presence of G418 and T7 and Sp6 promoters flanking the multiple cloning site. The 39 bases of the 5'-noncoding region, immediately upstream of end including the ATG translation initiation codon of Gαi2 and Gαi3 are selected for use as antisense probes to take advantage of the diversity of the nucleotide sequence in this region and to provide specificity (48% identity in the noncoding region vs >85% identity in the coding region; *see* **ref.** *13* and **Note 1**). The construction of vectors is performed with the use of standard techniques:

1. Complementary oligodeoxynucleotides are hybridized together and phosphorylated:
   a. Heat oligodeoxynucleotides Gαi2 or Gαi3 sense and antisense (10 nmol in 20 μL of water) for 5 min at 100°C, and then place on ice.
   b. Phosphorylate the double-strand DNA with T4 polynucleotide kinase (5 U): to 20 μL of hybridized complementary oligodeoxynucleotide in 59 μL of water, add 10 μL T4 kinase 10X buffer, 10 μL rATP (10 m*M*), and 1 μL T4 polynucleotide kinase (5-10 U). Incubate for 1 h at 37°C.
   c. Then perform phenol/chloroform (v/v) extraction, 75% ethanol precipitation, resolubilize DNA in 10 μL water, and store at 4°C until use.
2. Insert the double strand into the *Eco*RV cloning site of the polylinker of pcDNA3 expression vector:
   a. Digest the pcDNA3 expression vector with EcoRV restriction endonuclease: to 5 μg of pcDNA3 vector in 17 μL water, add 2 μL EcoRV 10X reaction buffer, 1 μL *Eco*RV restriction endonuclease (10 U). Incubate for 1 h at 37°C.
   b. Dephosphorylate the digested vector with alkaline phosphatase: to 20 μL of *Eco*RV-digested pcDNA3, add 5 μL dephosphorylation buffer (10X) and 5 μL alkaline phosphatase (5 U). Add water to make up 50 μL. Mix gently. Incubate at 37°C for 1 h.
   c. Then perform phenol/chloroform (v/v) extraction, 75% ethanol precipitation, and resolubilize pcDNA3 in 5 μL water.
   d. Set up ligation reaction: 2 μL double-strand DNA (0.2 pmol), 2 μL T4 DNA ligase 10X buffer, 1 μL pcDNA3 vector (100 ng). Add water to make up to 19 μL. Microcentrifuge. Vortex. Microcentrifuge. Add 1 μL T4 DNA ligase (1–3 Weiss units). Mix gently. Incubate at 16°C overnight.
3. Transformation into competent *E. coli* XL-1 cells and ampicillin selection:
   a. Thaw competent *E. coli* XL-1 cells gently at 4°C. Mix carefully, and pipet 50 μL into a precooled tube.
   b. Add 4 μL ligation reaction to cells. Tap tube and leave on ice for 30 min.
   c. Heat shock at 42°C for 50 s and incubate on ice for 2 min.
   d. Add 900 μL LB medium, and shake at 200–250 rpm for 1 h at 37°C.
   e. Plate 100 μL onto LB agar plates containing 0.05 mg/mL ampicillin.
   f. Incubate inverted at 37°C overnight in a humidified incubator.
   g. Select 5–10 colonies containing insert, and core from the dish using a pipet tip, then transfer to 30 mL LB with 0.05 mg/mL ampicillin, and grow at 37°C in a shaker (250 rpm) for 12–15 h.
4. DNA minipreparation (using a Nucleobond AX100 cartridge, Macherey-Nagel) and sequencing:
   a. Centrifuge the bacterial culture at 4000*g* for 15 min at 4°C.
   b. Resuspend the bacterial cell pellet carefully in 4 mL of buffer S1.

c. Add 4 mL of buffer S2 and mix the suspension gently by inverting tube.
d. Incubate the mixture at RT for 5 min.
e. Add 4 mL of buffer S3, mix gently until a homogenous suspension is formed, and then incubate on ice for 5 min.
f. Equilibrate the plasmid purification cartridge with 2 mL of buffer N2.
g. Centrifuge the suspension for 30 min at 20,000$g$ at 4°C, and then load the supernatant on the preequilibrated cartridge.
h. Wash the cartridge with 4 mL of buffer N3. Repeat this washing step.
i. Elute the plasmid DNA with 2 mL of buffer N5. Repeat this elution step.
j. Precipitate the purified plasmid DNA with 0.7 vol of isopropanol pre-equilibrated to RT, and centrifuge at high speed (>11,000$g$) at 4°C. Wash the DNA with 70% ethanol, dry briefly, and redissolve in water at 1 µg/µL.
k. Quantify the yield of DNA using a spectrophotometer by measuring the absorbance at 260 nm and 280 nm. The ratio of absorbance at 260 and 280 nm should exceed 1.8.

5. Plasmid inserts are sequenced by a dideoxy chain-termination method, using the commercially available kit SEQUENASE 2.0 kit (Amersham) and analyzed for the orientation of inserts (*see* **Note 2**).

## 3.3. Transfection of Antisense Gαi Subunit Expression Vector

Cl.10 cells are transfected by electroporation using a gene pulser (Electro-porator II, Invitrogen).

1. Remove $5 \times 10^6$ exponentially growing cells from the plastic flask as described above, and preincubate them on ice for 5 min with 20 µg of pcDNA3 plasmid encoding Gαi2 or Gαi3 antisense or without cDNA insert and 20 µg of salmon sperm DNA carrier in 500 µL cold DMEM/Ham's F-12 medium with 100 IU/mL penicillin and 100 µg/mL streptomycin.
2. Perform electroporation at 330 V and 500 µF in an electroporation cuvette (0.4 cm).
3. Keep cells on ice for 5 min after electroporation, and add 5 mL of standard culture medium. Transfer to a 25-cm² plastic culture flask.
4. Seed transfected Cl.10 cells 48 h after electroporation, and select them by addition of geneticin (G418) to a final concentration of 400 µg/mL for 3 weeks.
5. Cl.10 cells that are resistant to geneticin are subsequently cloned by limiting dilution as described above and characterized for their Gαi content by Western blotting as described below (*see* **Note 3**).

## 3.4. Preparation of Particulate Fraction of Cultured Cells

1. Wash three times with 10 mL of 0.13 $M$ PBS (pH 7.2) the control and transfected Cl.10 cells grown in 75-cm² plastic culture flasks for 13 days (as described above), treated or not treated with pertussis toxin.
2. Harvest washed cells in 10 mL 0.13 $M$ PBS using a rubber policeman, and centrifuge at 2000$g$ for 5 min at 4°C.

3. Expose the cell pellet for 30 min to 5 mL hypoosmotic 5 m$M$ HEPES buffer, pH 7.4.
4. Centrifuge aliquots (500 µg of proteins) of cell suspensions at 20,000$g$ for 15 min at 4°C, wash with 1 mL of 20 m$M$ HEPES buffer, pH 7.4, pellet, and store at − 80°C until use.

This particulate fraction from cell homogenates is referred to as membrane preparation.

### 3.5. Immunoblotting of Gαi Subunit of Gi-Proteins

Gi-proteins are involved in the coupling of PYY receptors to biological events in Cl.10 cells. Western blot analysis indicated the presence of the Gαi2 ($M_r$ 39,000) and Gαi3 ($M_r$ 42,000) subunits, but not the Gαi1 subunit in Cl.10 cell membranes. To analyze the extinction of Gαi-subunit protein expressed in control Cl.10 cells and transfected Cl.10 cells, we carry out gel electrophoresis followed by Western blotting using specific antibodies against Gαi1/Gαi2 subunits (AS7) or Gαi3/Gαo (EC2).

1. Typically, for the analysis of Gαi protein samples, a 10% polyacrylamide gel is run. The following parameters are described for a Hoefer SE250 gel kit (Hoefer, San Francisco, CA). For the running-gel solution, mix 5 mL of 40% acrylamide stock with 5 mL of 1.5 $M$ Tris base, pH 8.8 containing 0.4% SDS and 10 mL water. Add 50 µL of a fresh solution of 10% ammonium persulfate and 25 µL of TEMED to polymerize the gel.
2. Prepare a 5% stacking gel (1.25 mL stock 40% acrylamide, 2.5 mL 0.5 $M$ Tris base, pH 6.8, containing 0.4% SDS and 6.25 mL water). Polymerize the gel with the addition of 50 µL 10% ammonium persulfate and 25 µL TEMED.
3. Pour the running gel in between the glass plates of the gel apparatus, leaving 2 cm clear at the top. Carefully layer butanol on the top, and leave the gel to polymerize. (At this stage, gels can be left for several days if wrapped in clingfilm at 4°C). To use, pour the butanol off, add the stacking gel, and insert the gel-comb of choice.
4. Boil the solubilized protein preparations in sample buffer (50 µg proteins/40 µL sample buffer) for 5 min, and load the gel.
5. Electrophorese the samples at 30 mA in the running buffer until the blue dye has reached the bottom of the gel. Gels are calibrated with molecular weight marker proteins: myosin (200,000), phosphorylase b (97,000), BSA (68,000), ovalbumin (43,000), and carbonic anhydrase (29,000).
6. Remove the gel from the plates, and place in 50-mL transfer buffer for 15 min. At the same time, soak six pieces of Whatman 3MM paper in the transfer buffer. (These paper sheets should have been cut to the same size as the gel.)
7. Assemble for transfer in the following order: three pieces of 3MM paper, the cellulose nitrate membrane, the gel, and finally three pieces of 3MM paper. These are placed into the semi-dry electrophoretic transfer cell (Bio-Rad) with the membrane toward the anode side.

8. Transfer for 1 h at 20 V.
9. Remove the membrane, and incubate for 1 h (with gentle shaking) in 10 mL antibody buffer A. Rinse for 15 min with 10 mL antibody buffer B.
10. Pour off the solution, and add 10 mL of primary antibody (AS7 diluted 1:1,000 or EC2 diluted 1:1,000 in antibody buffer B). Shake slowly for 90 min at RT at 4°C overnight.
11. Wash three times for 15 min with 10 mL antibody buffer A, and rinse one time with 10 mL antibody buffer B.
12. Add $^{125}$I-labeled goat antibodies to rabbit IgG in 10 mL antibody buffer B and incubate for 1 h at RT with gentle shaking.
13. Wash three times for 15 min with 10 mL antibody buffer A, and rinse one time with 0.13 *M* PBS⁻.
14. Dry cellulose nitrate membrane prior to autoradiography.
15. Autoradiograms of the dried immunoblots are scanned with a Macintosh Onescanner densitometer to estimate the relative amount of Gαi subunits expressed in transfected clone cells.

This Western blot revealed a Cl.10/Gαi2 clone, among 20 G418-resistant clones, which displayed a drastic decrease (>90%) in expression of Gαi2 without changes of Gαi3, and a Cl.10/Gαi3 clone, among 13 G418-resistant clones, which displayed a drastic decrease (>80%) in expression of Gαi3 without changes in Gαi2 compared with Cl.10 cells transfected with pcDNA3 vector alone (**Fig. 1A**).

These two clones are selected to study the Gi-coupled PYY receptors by PYY binding, cAMP measurement, or thymidine incorporation.

### 3.6. PYY Binding to Membrane Receptors

Competitive inhibition of [$^{125}$I]PYY binding by unlabeled PYY is performed on membranes prepared from control Cl.10 cells, Cl.10/Gαi2 clone cells, and Cl.10/Gαi3 clone cells. All binding data are analyzed using the LIGAND computer program developed by Munson and Rodbard *(14)*.

1. Prepare and purify iodinated PYY (synthetic PYY is radiolabeled with [$^{125}$I]Na using the chloramine T method):
   a. 14 µg of PYY diluted in phosphate buffer 0.3 *M*, pH 7.4 is radiolabeled for 30 s at RT with 1 mCi (5 µL) [$^{125}$I]Na in the presence of 5 µL of chloramine T (1 mg/mL).
   b. Stop the reaction by addition of 5 µL of $Na_2S_2O_5$ (2 mg/mL).
   c. Perform the radiolabeled PYY purification using reverse-phase high-performance liquid chromatography in isocratic conditions (1 ml/min, 31% acetonitrile in 0.1% trifluoroacetic acid). Two principal tracers are obtained in these conditions: [$^{125}$I-Tyr$^1$]monoiodo-PYY and [$^{125}$I-Tyr$^{36}$]monoiodo-PYY.
   d. Fractions containing the radiolabeled tracer are stored at –20°C in 0.1 *M* acetic acid containing 0.1% BSA.

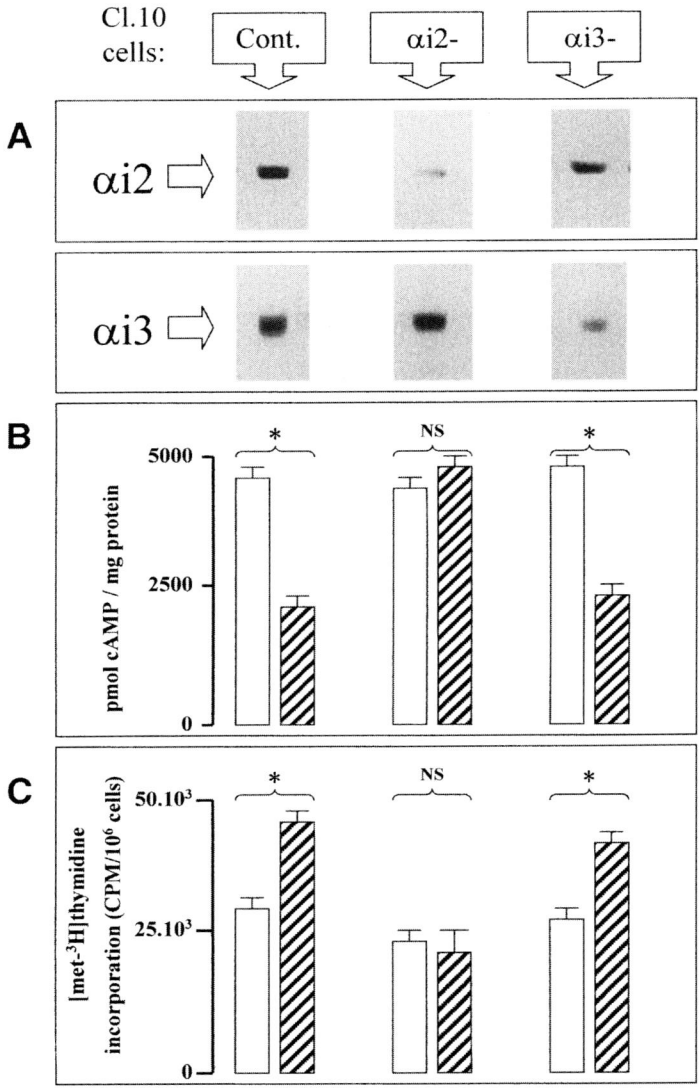

Fig. 1. Analysis of coupling of peptide YY receptor to Gi-protein in renal proximal tubule Cl.10 cells. (**A**) Western blotting of Gαi2 and Gαi3 in Cl.10 control cells, Cl.10/Gαi2- (Gαi2-), and Cl.10/Gαi3- (Gαi3-) cells. Cl.10/Gαi2- cells display down-regulation in the expression of Gαi2 protein (>90%) without modification in Gαi3 protein content as compared with Cl.10 control cells. In contrast, Cl.10/Gαi3- cells display a large reduction in the expression of Gαi3 protein (>80%) with no alteration in Gαi2 protein expression. (**B**) Effect of PYY on forskolin-stimulated cAMP levels in Cl.10 control cells, Cl.10/Gαi2-, and Cl.10/Gαi3- cells. Cells are incubated with 10 μ*M*

e. The [$^{125}$I-Tyr$^{36}$]monoiodo-PYY (referred to as [$^{125}$I]PYY below) shows a much higher level of specific binding to Cl.10 cell membranes than [$^{125}$I-Tyr$^{1}$]monoiodo-PYY and is therefore used for all binding experimental procedures. [$^{125}$I]PYY is used during the first month following labeling without a significant loss of binding activity.

2. Incubate Cl.10 cell membranes (200 μg protein/mL) for 90 min at 30°C in 250 μL of binding incubation buffer containing 0.05 n$M$ [$^{125}$I]PYY (2200 Ci/mmol) with or without increasing concentrations of unlabeled PYY ($10^{-6}$–$10^{-11}$ $M$).

3. At the end of the incubation, mix 150-μL aliquots of membranes with 150 μL ice-cold binding incubation buffer.

4. Bound and free peptides are separated by centrifugation at 20,000$g$ for 10 min at 4°C, and membrane pellets are washed twice with 10% (w/v) sucrose in 20 m$M$ HEPES buffer, pH 7.4.

5. Count the radioactivity with a γ-counter. The nonspecific binding represents about 2% of total radioactivity.

6. Measure proteins using a protein assay kit (Bio-Rad) based on the method of Bradford *(15)* with BSA as a standard.

Scatchard analysis of PYY receptors to Cl.10/Gαi2- transfected cells, compared to untransfected Cl.10 cells, showed an increase in the dissociation constant of the receptor (5.3 n$M$ vs 0.6 n$M$). This increase is identical to that observed in pertussis toxin-treated untransfected cells. The dissociation constant of PYY receptors expressed in Cl.10/Gαi3- transfected cells is identical to that observed in untransfected Cl.10 cells. These findings support a coupling of PYY receptor to Gαi2 protein.

### 3.7. cAMP Measurement

As PYY inhibited cAMP production via a pertussis toxin-sensitive Gi-protein in the mouse proximal tubule cell line PKSV-PCT Cl.10, we further investigated the influence of expression of Gαi2 or Gαi3 antisense RNA on

---

forskolin in the absence (open bars) or presence (hatched bars) of 1 μ$M$ PYY. Compared with Cl.10 control cells, Cl.10/Gαi2- cells exhibit failure of PYY to inhibit cAMP levels. The PYY-mediated-inhibition of cAMP in Cl.10/Gαi3- cells is identical to that observed in Cl.10 control cells. **(C)** Effect of PYY on [methyl-$^{3}$H]thymidine incorporation into DNA in Cl.10 control cells, Cl.10/Gαi2- (Gαi2-) cells, and Cl.10/Gαi3- (Gαi3-) cells. Two days after seeding, cells are cultured in fetal calf serum-deprived medium in the presence (hatched bars) or absence (open bars) of 0.1 μ$M$ PYY for 18 h. Compared with Cl.10 control cells, Cl.10/Gαi2- cells exhibit the failure of PYY to stimulate [methyl-$^{3}$H]Thymidine incorporation into DNA. The PYY-mediated stimulation of [methyl-$^{3}$H]Thymidine incorporation into DNA in Cl.10/Gαi3- cells is identical to that observed in Cl.10 control cells. *$p < 0.005$ vs control without PYY; NS, nonsignificant.

PYY-inhibited cAMP production. Cellular cAMP content is assayed in basal condition, in the presence of 10 $\mu M$ forskolin, which stimulates adenylyl cyclase, and in the presence of 10 $\mu M$ forskolin and 0.1 $\mu M$ PYY.

1. Seed control or transfected Cl.10 cells in 12-well trays, and culture them with 1 mL of standard culture medium for 5 days. Culture medium is changed every 2 days.
2. Incubate cells in 12-well trays for 1 h at 37°C with 1 mL of DMEM/Ham's F-12 without fetal calf serum.
3. Then incubate cells in 12-well trays in the presence or absence of 10 $\mu M$ forskolin in 1 mL of DMEM/Ham's F-12 containing 2% (w/v) BSA, 0.1% (w/v) bacitracin, and 0.2 m$M$ 3-isobutyl-1-methylxanthine (which inhibits cAMP degradation by phosphodiesterase) without or with 0.1 $\mu M$ PYY for 40 min at 37°C.
4. At the end of incubation, remove the medium rapidly, and wash the cells in 1 mL 0.13 $M$ PBS⁻, pH 7.2.
5. Add 1 mL of ice-cold 95% ethanol/5% formic acid, and harvest precipitated cells using a rubber policeman. Centrifuge at 4000$g$ for 15 min at 4°C.
6. Evaporate supernatants to dryness.
7. Determine cAMP content by radioimmunoassay.
8. Make cell protein determinations in parallel wells. Report data as pmol of cAMP per mg protein.
9. Calculate statistical significance between groups by Student's $t$-test.

PYY does not inhibit cAMP production in Cl.10/G$\alpha$i2-transfected cells. The PYY-mediated inhibition of cAMP measured in Cl.10/G$\alpha$i3-transfected cells is identical to that observed in untransfected Cl.10 cells. These findings support a coupling of PYY receptor to adenylyl cyclase via the G$\alpha$i2 protein (**Fig. 1B**).

### 3.8. [Methyl-³H]Thymidine Incorporation

As PYY stimulated cell growth in mouse proximal tubule PKSV-PCT Cl.10 cells, cell growth kinetics are estimated by [methyl-³H]Thymidine (25 Ci/mmol) incorporation in transfected Cl.10 cells and control Cl.10 cells.

1. Harvest cells with 2.7 m$M$ EDTA without trypsin. (Membrane proteins are not hydrolyzed by EDTA treatment.)
2. Seed cells (15,000 cells/dish) in 12-well trays, and culture for 3 days in standard culture medium.
3. Incubate cells in the absence or presence of 0.1 $\mu M$ PYY for 18 h in serum- and growth factor-free culture medium.
4. Incubate cells for 6 h with [methyl-³H]thymidine (0.5 mCi/well).
5. Rinse cells three times with 1 mL ice-cold 0.13 $M$ PBS.
6. Add 1 mL 5% trichloroacetic acid for 30 min at 4°C.
7. Remove trichloroacetic acid, and incubate the cells for 30 min at 37°C with 1 mL 0.3 $N$ NaOH.

8. After neutralization with acetic acid, measure the radioactivity in the cell extracts by scintillation counting. Express the results as cpm per $10^6$ cells.
9. Count cells before addition of trichloroacetic acid.

PYY does not stimulate [methyl-$^3$H]thymidine incorporation into DNA in Cl.10/Gαi2- transfected cells. The PYY-mediated stimulation of [methyl-$^3$H]thymidine incorporation into DNA measured in Cl.10/Gαi3-transfected cells is identical to that observed in untransfected Cl.10 cells.

In conclusion, antisense RNA technology studies indicate that PYY receptors are coupled with a strict specificity to Gαi2 protein in the proximal tubule Cl.10 cell clone and that Gαi2 is responsible for PYY receptor-mediated inhibition of adenylyl cyclase and stimulation of cell growth. These findings further document the mechanism of PYY receptor-mediated responses in epithelial cells *(16)* (**Fig. 1C**).

## 4. Notes

1. The present investigation, which takes advantage of the powerful antisense RNA technology, is the first to demonstrate the coupling of PYY receptors to the Gαi2 protein. This is possible because selection of 39 bases of the 5'-noncoding region of Gαi2 or Gαi3 for use as antisense templates provided the necessary nucleotide sequence specificity, e.g., only 48% identity in this region, whereas selection of templates in the coding region with >85% identity would probably have failed to ensure such specificity.
2. The double-stranded Gαi2 or Gαi3 DNA ligates in either orientation to the *Eco*RV cloning site of the polylinker of pcDNA3. Thus it is very important to verify the insert's orientation by DNA sequencing before transfection.
3. After transfection with the pcDNA3 antisense Gαi2, selection, and cloning, 20 clones are isolated, and their level of expression of Gαi2 is characterized by Western blot. Among 20 G418-resistant clones, 17 clones did not display downregulation in the expression of Gαi2 protein. This fact suggest the importance of subcloning stably transfected cells.

## References

1. Laburthe, M., Chenut, B., Rouyer-Fessard, C., Tatemoto, K., Couvineau, A., Servin, A., et al. (1986) Interaction of peptide YY with rat intestinal epithelial plasma membranes : binding of the radioionated peptide. *Endocrinology* **118,** 1910–1917.
2. Voisin, T., Bens, M., Cluzeaud, F., Vandewalle, A., and Laburthe, M. (1993) Peptide YY receptors in the proximal tubule PKSV-PCT cell line derived from transgenic mice : relation with cell growth. *J. Biol. Chem.* **268,** 20,547–20,554.
3. Laburthe, M. (1990) Peptide YY and neuropeptide Y in the gut, availability, biological actions, and receptors. *Trends Endocrinol. Metab.* **1,** 168–174.
4. Servin, A. L., Rouyer-Fessard, C., Balasubramaniam, A., Saint-Pierre, S., and Laburthe, M. (1989) Peptide-YY and neuropeptide Y inhibit vasoactive intestinal

peptide-stimulated adenosine 3',5'-monophosphate production in rat small intestine: structure requirements of peptides for interacting with peptide-YY-preferring receptors. *Endocrinology* **124,** 692–700.

5. Castan, I., Valet, P., Voisin, T., Quideau, N., Laburthe, M., and Lafontan, M. (1992) Identification and functional studies of a specific peptide YY-preferring receptor in dog adipocytes. *Endocrinology* **131,** 1970–1976.

6. Voisin, T., Rouyer-Fessard, C., and Laburthe, M. (1990) Distribution of common peptide YY-neuropeptide Y receptor along rat intestinal villus-crypt axis. *Am. J. Physiol.* **255,** G113-G120.

7. Balasubramaniam, A., Cox, H. M., Voisin, T., Laburthe, M., Stein, M., and Fischer, J. E. (1993) Structure-activity studies of peptide YY(22–36): N-α-Ac-[Phe$^{27}$]PYY(22–36), a potent antisecretory peptide in rat jejunum. *Peptides* **14,** 1011–1016

8. Voisin, T., Couvineau, A., Rouyer-Fessard, C., and Laburthe, M. (1991) Solubilization and hydrodynamic properties of active peptide YY receptor from rat jejunal crypts: characterization as a Mr 44,000 glycoprotein. *J. Biol. Chem.* **266,** 10,762–10,767

9. Cartier, N., Lacave, R., Vallet, V., Hagege, J., Hellio, R., Robine, S., et al. (1993) Establishment of renal proximal tubule cell lines by targeted oncogenesis in transgenic mice using the L-pyruvate kinase-SV40 (T) antigen hybrid gene. *J. Cell Sci.* **104,** 695–704

10. Lacave, R., Bens, M., Cartier, N., Vallet, V., Robine, S., Pringault, E., et al. (1993) Functional properties of proximal tubule cell lines derived from transgenic mice harboring L-pyruvate kinase-SV40 (T) antigen hybrid gene. *J. Cell Sci.* **104,** 705–712.

11. Taussig, R. and Gilman, A. G. (1995) Mammalian membrane-bound adenylyl cyclases. *J. Biol. Chem.* **270,** 1–4.

12. Pouysségur, J (1990) G proteins in growth factor action, in *G Proteins*, (Iyengar, R., ed.), Academic, NY, pp. 555–570.

13. Moxham, C. M., Hod, Y., and Malbon, C. C. (1993) Induction of G alpha i2-specific antisense RNA in vivo inhibits neonatal growth. *Science* **260,** 991–995.

14. Munson, P. J. and Rodbard, D. (1980) Ligand: a versatile computerized approach for characterization of ligand-binding systems. *Anal. Biochem.* **197,** 220–239.

15. Bradford, M. (1976) A rapid and sensitive method for the quantitation of microgram quantities of protein utilizing the principle of protein-dye binding. *Anal. Biochem.* **72,** 248–254.

16. Voisin, T., Lorinet, A. M., Maoret, J. J., Couvineau, A., and Laburthe, M. (1996) Gαi RNA antisense expression demonstrates the exclusive coupling of peptide YY receptors to Gi2 protein in renal proximal tubule cells. *J. Biol. Chem.* **271,** 574–580.

# 11

## NPY Y5 Receptor Subtype

*Pharmacological Characterization*
*with Antisense Oligodeoxynucleotide Screening Strategy*

### Andrea O. Schaffhauser, Alain Stricker-Krongrad, and Karl G. Hofbauer

### 1. Introduction

Neuropeptide Y (NPY), peptide YY (PYY), and pancreatic polypeptide (PP) are endogenous 36-amino acid peptides belonging to the same family. Different receptor subtypes have been identified in the NPY/PYY/PP family *(1)*, and mammalian subtypes are now classified collectively as NPY receptors, several of which (namely Y1, Y2, Y4, Y5, and Y6) have been cloned *(2–10)*. Based on studies using full-length peptides, C- or N-terminal fragments, and modified peptides of the NPY/PYY/PP family *(11,12)*, the NPY Y5 receptor subtype was proposed as a mediator of NPY-induced feeding. At the time when the NPY Y5 receptor was cloned, neither NPY Y5 antagonists nor NPY Y5 knockout mice were available. Therefore we used an antisense oligodeoxynucleotide (antisense ODN) strategy to assess the proposed role of the NPY Y5 receptor subtype in the regulation of feeding.

The essentials of antisense technology have been reviewed previously by Gold *(13)*. Thus only a brief summary is presented here. The potency of antisense ODNs depends on the target site at the mRNA, where three regions are most suitable, namely, the 5'-cap, the AUG translation initiation codon, and the 3' untranslated region. Combination of several ODN sequences may result in increased activity. Furthermore, the length as well as chemical modifications of the natural phosphodiester ODNs are crucial for the efficacy of antisense ODNs. The use of at least two ODN control sequences, such as sense,

From: *Methods in Molecular Biology, vol. 153: Neuropeptide Y Protocols*
Edited by: A. Balasubramaniam © Humana Press Inc., Totowa, NJ

scrambled, or mismatched ODNs, is recommended to rule out nonspecific effects and toxicity.

Bearing this in mind, we designed seven different 20-base phosphodiester antisense ODN sequences, four targeted to the start and three targeted to the stop codon region of the NPY Y5 mRNA. Although low nuclease activity is found in the cerebrospinal fluid (CSF), Whitesell et al. *(14)* showed that modifications to protect at least the 3' base linkage from exonuclease attack are required to stabilize ODNs in the CSF. Therefore, phosphothioate-terminal-protected derivatives were also synthesized. It has been suggested by Wahlestedt *(15)* that treatment of at least 2 days is needed for the effective reduction of G-protein-coupled receptor density. For this reason antisense ODNs were applied in an in vivo screening profiling experiment over a period of 2 days. Based on the assumption that the NPY Y5 receptor subtype was the NPY "feeding" receptor, basal and NPY-induced food intake following antisense treatment were chosen as selection criteria for active sequences. Furthermore, the hypothesis that a combination of two antisense sequences would exceed the effects of a single antisense sequence was also tested.

Experiments with control sequences were done to prove specificity of the antisense effect. Eventually, two start and one stop codon NPY Y5 antisense sequences that had been active in both in vivo screening models were selected. These antisense sequences, as well as their mismatched and sense control sequences were synthesized as phosphothioate-terminal-protected derivatives to increase their stability against nuclease attack. If the observed effect was specific, no effect should be observed with other orexigenic peptides. To investigate this, galanin instead of NPY was injected after pretreatment with NPY Y5 antisense sequences. Recent data published by Monia et al. *(16)* illustrated the importance of a combination of in vitro and in vivo experiments to demonstrate antisense activity. In our case, the documentation of a specific antisense effect at the cellular level was incomplete for several reasons. The lack of a suitable NPY Y5 ligand precluded the ex vivo determination of NPY Y5 receptors in brain tissue. Additionally, no studies had been performed to elucidate the cellular uptake, distribution, and stability of the different antisense derivatives.

Nonetheless, we investigated the mechanism of action of three antisense sequences that were most active in the in vivo screening models using two approaches. The purpose of these studies was to define whether inhibition of protein synthesis by antisense ODNs was a consequence of translation arrest, inhibition of RNA processing, or RNase H cleavage of the DNA-mRNA duplex *(17,18)*. If translation arrest was the underlying mechanism of action, NPY Y5 receptor densities should be decreased after antisense treatment. Therefore, we determined NPY Y5 receptor binding after antisense treatment in two different

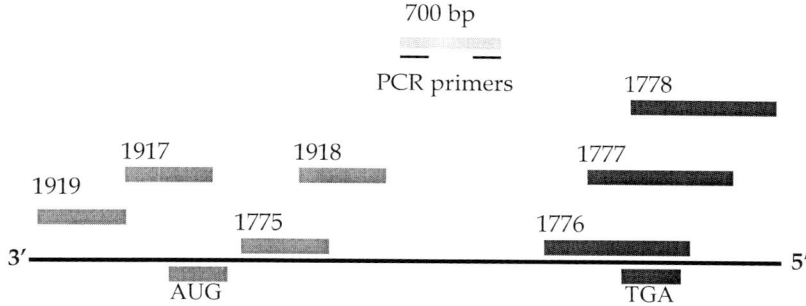

Fig. 1. NPY Y5 mRNA.

cell lines expressing the NPY Y5 receptor subtype. To determine whether RNase H cleavage of the DNA-mRNA duplex was involved, subsequent reverse transcription-polymerase chain reaction (RT-PCR) was performed on rat hypothalamic tissue after NPY Y5 antisense treatment.

## 2. Materials

### 2.1. Antisense ODN Screening Strategy

The schematic map (shown in **Fig. 1**) of NPY Y5 mRNA illustrates the positioning of the different NPY Y5 start and stop codon antisense ODNs and PCR primers:

1. The following antisense ODN sequences are used for the rat Y5 receptor mRNA *(5)*, targeting the *start codon*:

    | | |
    |---|---|
    | 1917 ODN | 5' AGA GGA CGT CCA TTA GCA GC 3' |
    | 1918 ODN | 5' CTT AAA CTC CAT ACT AGA ATC CTG 3' |
    | 1919 ODN | 5' CCA TTA GCA GCT GCA GAG AGA ACG 3' |
    | 1775 ODN | 5' TGG TGG AAG AAG AGG ACG TC 3' |

    Phosphodiester ODNs as well as phosphothioate-terminal-protected ODN sequences are tested.

2. The following antisense ODN sequences are used for the rat Y5 receptor mRNA *(5)*, targeting the *stop codon*:

    | | |
    |---|---|
    | 1776 ODN | 5' TCA TGA CAT GTG TAG GCA GT 3' |
    | 1777 ODN | 5' CAG AGA GAA TCA TGA CAT GT 3' |
    | 1778 ODN | 5' TTG GTG CAC AGA GAG AAT CA 3' |

    Phosphodiester ODNs as well as phosphothioate-terminal-protected ODN sequences are tested.

3. Phosphothioate-terminal protected control sequences are used for the antisense ODNs 1917, 1918 and 1776.

    | | |
    |---|---|
    | 1917 mismatched ODN | 5' AGA GTA CGG CCA CTA GCA GT 3' |
    | 1917 sense ODN | 5' GCT GCT AAT GGA CGT CCT CT 3' |
    | 1918 mismatched ODN | 5' CTT AAA CTT CAC ATT AGA ACC CTG 3' |

| 1918 sense ODN | 5' CAG GAT TCT AGT ATG GAG TTT AAG 3' |
| 1776 mismatched ODN | 5' TAA TGA CAG GTG TCG GCA TT 3' |
| 1776 sense ODN | 5' ACT GCC TAC ACA TGT CAT GA 3' |

*See* **Note 4.3.**

## 2.2. In Vivo Studies

1. Saline: i.e., 0.9% NaCl is used as vehicle.
2. Porcine NPY (300 pmol/5 µL) is dissolved in saline.
3. Galanin (2000 pmol/5 µL) is dissolved in saline.
4. The solubility of ODNs is critical and varies with the protection used. The ODNs are dissolved in saline (50 µg/10 µL), and the concentration is confirmed for each ODN by photometry (1 $OD/mL_{260nm}$ = 37 µg/mL ODN) prior to use.

## 2.3. Protocol for Ex Vivo NPY Y5 mRNA Determination

### 2.3.1. ODNs for RT-PCR (see **Note 4.3.**)

The following ODN primers are used to amplify the rat Y5 cDNAs *(5)*:

| Rat Y5 antisense ODN | 5' GGT CAG TCT GTA GAA AAC ACT TCG AGA TC 3' |
| Rat Y5 sense ODN | 5' TGG ATC AGT GGA TGT TTG GCA AAG CC 3' |
| β-actin ODN | 5' CAC CCA CAC TGT GCC CAT CTA TG 3' |
| β-actinR ODN | 5' AGA AGC ATT TGC GGT GCA TGG 3' |

### 2.3.2. Quantification of NPY Y5 mRNA by RT-PCR

1. 1.5% agarose gel: 1.5 g agarose, 5 µL ethidium bromide (10 mg/mL), 10 mL 10X TBE buffer, 90 mL $H_2O$.
2. RT buffer for Y5 mRNA quantification: 2 µL 25 m$M$ $MgCl_2$, 1 µL 10X PCR buffer (500 m$M$ KCl, 100 m$M$ Tris-HCl, pH 8.3), 0.5 µL 20 m$M$ dNTP mix, 0.5 µL poly dT 15 primer (33 µg/µL), 0.3 µL RNase inhibitor 40,000 U/mL, 0.125 µL SuperScript II RT 200 U/µL, 1 µL synthetic Y5 mimic at different concentrations (12.5, 6.5, 3, 1.5, 0.75 fg/µL).
3. RT buffer for β-actin mRNA quantification: 2 µL 25 m$M$ $MgCl_2$, 1 µL 10X PCR buffer (500 m$M$ KCl, 100 m$M$ Tris-HCl, pH 8.3), 0.5 µL 20 m$M$ dNTP mix, 0.5 µL poly(dT) 15 primer (33 µg/µL), 0.3 µL RNase inhibitor 40,000 U/mL, 0.125 µL SuperScript II RT 200 U/µL.
4. Loading buffer: 0.1 M EDTA, 1% SDS, 20% glycerol, 0.05% bromo-phenol-blue.
5. Lysis buffer: 4 M urea, 100 m$M$ HEPES-KOH, pH 7.5, 180 m$M$ NaCl, 1% sodium dodecyl sulfate (SDS), 5 m$M$ dithiothreitol (DTT) and 400 µg/mL protein kinase K.
6. PCR buffer for Y5 mRNA quantification: 2 µL 25 m$M$ $MgCl_2$, 4 µL 10X PCR buffer (500 m$M$ KCl, 100 m$M$ Tris-HCl, pH 8.3), 33 µL $H_2O$, 1 µL sense primer (25 nmol/mL), 1 µL antisense primer (25 nmol/mL), 0.25 µL taq DNA polymerase.

7. PCR buffer for β-actin mRNA quantification: 2 μL 25 m$M$ MgCl$_2$, 4 μL 10X PCR buffer (500 m$M$ KCl, 100 m$M$ Tris-HCl, pH 8.3), 33 μL H$_2$O, 1 μL sense primer (25 nmol/mL), 1 μL antisense primer (25 nmol/mL), 1 μL β-actin mimic at different concentrations (6, 3, 1.5, 0.75, 0.375 fg/μL), 0.25 μL taq DNA polymerase.
8. 10X TBE buffer, pH 8.0: Tris base 108 g, boric acid 55 g, 0.5 M EDTA 40 mL, H$_2$O to 1000 mL.

## 2.4. In Vitro Studies: Protocols for Using a Surrogate Cell Model

### 2.4.1. Cell Culture

1. Cell culture medium: Dulbecco's modified Eagle's medium (DMEM) with 10% bovine calf serum (BCS), 2 m$M$ L-glutamine, 100 UI/mL penicillin/100 μg/mL streptomycin, 300 μg/mL geniticin.
2. Isotonic solution: isoton II.
3. Coating reagent: 0.1% poly-D-lysin-diluted H$_2$O.
4. HEK-293 cells: human embryonic kidney cell line stably transfected with the rat Y5 receptor.
5. LMTK cells: mouse fibroblast cell line stably transfected with the rat Y5 receptor.
6. Serum reduced medium: Opti-MEM®.
7. Transfection Reagent: DOTAP®.
8. Trypsin-EDTA: 0.5 g trypsin (1:250) and 0.2 g EDTA.
9. Washing solution: phosphate-buffered saline (PBS), pH 7.4.

### 2.4.2. Radioligand Binding Assay in Cell Culture
(See **Note 4.3.**)

1. Radioligand: [$^{125}$I-Pro34]PYY (human; specific activity about 2130 Ci/mmol, 25 p$M$ for HEK-293 cells and 150 p$M$ for LMTK cells), diluted in incubation medium.
2. Competitor: [Pro34]hPYY; 1 μ$M$.
3. Washing solution: PBS, pH 7.4.
4. Incubation medium: DMEM with 4 m$M$ MgCl$_2$, 4 m$M$ CaCl$_2$, 10 m$M$ HEPES, 0.1% bacitracin, and 50 μ$M$ phenylmethylsulfonyl fluoride (PMSF).

## 2.5. Reagents

1. β-actin mimic: Clontech, Palo Alto, CA.
2. β-actin primer: Microsynth, Balgach, Switzerland.
3. Dental cement: KETAC-CEM® APLICAP®, ESPE, Seefeld, Germany.
4. dNTP mix: Amersham Pharmacia Biotech, Uppsala, Sweden.
5. DOTAP: Roche, Basle, Switzerland.
6. Galanin: Neosystem, Strasbourg, France.
7. $^{125}$I-Pro$^{34}$]hPYY: Anawa, Wangen, Switzerland.
8. *Not*I/*Sac*I restriction sites: Roche, Basle, Switzerland.
9. OligodT: Amersham.
10. ODNs: Microsynth.

11. pNPY: Neosystem.
12. [Pro$^{34}$]hPYY: Saxon Biochemicals, Hannover, Germany.
13. QuickPrep Micro® mRNA: Amersham purification kit.
14. Rat chow diet: Nafag®, Gossau, Switzerland.
15. Y5 mimic: Clontech.
16. Y5 primer: Microsynth.

## 2.6. Statistics

The results are expressed as means ± SEM. Comparisons between experimental groups were done with Fisher's analysis of variance followed by Student's $t$-test (PLSD). Probability values <0.05 (two-tailed) are considered significant.

## 3. Methods

### 3.1. Antisense ODN Screening Strategy (see Note 4.5.)

The following flowchart illustrates the antisense ODN screening strategy that used antisense ODNs as pharmacological tools to characterize the NPY Y5 receptor.

In vivo model
    Primer design/RT-PCR probe design
    Antisense ODN design
    In vivo antisense ODN screening in rats
        Basal food intake
        Response to 300 pmol NPY intracerebroventricular (icv)
        Use of control sequences
        Use of galanin as control for specificity
    Ex vivo RNA quantification
        RT-PCR

In vitro model
    Protein quantification
        Receptor binding assay

### 3.2. Protocols for In Vivo Studies

*Intracerebroventricular cannulation*

1. Adult male Sprague-Dawley rats, weighing 180–280 g, are individually housed in stainless steel cages (22 ± 2°C; 12:12 h light/dark schedule, lights off at 6 PM). Tap water and rat chow diet are available ad libitum.
2. Rats are anaesthetized with pentobarbital 50 mg/kg ip and stereotaxically implanted with chronic guide cannulae (ED 0.6 mm, ID 0.35 mm, L 8.0 mm; Microfil, Renens, Switzerland) aimed at the right lateral ventricle.

3. Stereotaxic coordinates used, with the incisor bar set −2.0 mm below interaural line, are −0.8 mm anterior to bregma; +1.3 mm lateral to bregma according to coordinates from Paxinos and Watson *(19)*.
4. The guide cannula is placed on the dura and secured with screws and dental cement.
5. A stainless steel stylet (OD 0.3 mm), 1 mm longer than the guide cannula, is used to occlude the guide cannula during recovery and between injections.
6. Rats are treated with an analgesic (buprenorphin 0.03 mg/kg BW ip) 4–6 h after surgery.
7. Rats are tested for their feeding response to a 300-pmol pNPY icv injection to confirm correct placement of cannulae. Only rats showing a food intake above 1 g following the first hour post icv injection of NPY are used in the feeding studies.

*Acute icv injections*

1. After 1 week of postoperative recovery, rats of the same body weight (260–280 g) are matched for the studies. Rats are kept in individual cages throughout the experimental procedure. They can move freely and have free access to water and preweighed food.
2. The stylets are removed and replaced by injection cannulae (ED 0.3 mm, ID 0.1 mm, L 30.0 mm) that protrude 3.8 mm below the guide cannula, aiming at the right lateral ventricle. The injection cannula is fixed to a guide cannula at 11.8 mm, and the upper end is connected via polyethylene tubing to a microsyringe (Hamilton, Bonaduz, Switzerland).
3. Injections are delivered at a rate of 5 μL/30 s and 10 μL/60 s, and the injection cannula is kept in position for 30 s after the injection to avoid any reflux.
4. The injection cannula is withdrawn, and the guide cannula is occluded with the stylet.

*Hypothalamus sampling*

1. For ex vivo experiments, rats are anesthetized at the end of the experiment by carbon dioxide inhalation and killed by decapitation.
2. The brain is quickly removed and placed in a brain matrix, with the ventral surface facing up.
3. A 2-mm-long coronal section is made with the caudal optic chiasma as the anterior boundary, and the section is laid on a plate with the rostral surface facing up.
4. The hypothalamus is sampled using the top of the third ventricle and the lateral hypothalamic sulci as boundaries.
5. The collected brain regions are immersed in liquid nitrogen and stored at −80°C until used.

## 3.2.1. Antisense ODNs Design

Using a multitargeting approach, we designed several unmodified phospho-diester and phosphothioate-terminal-protected derivatives targeting different sites of the published NPY Y5 receptor *(5)*.

1. Four of these antisense ODNs (1917, 1918, 1919, and 1775 antisense ODNs) are targeted toward the start codon.
2. Three antisense ODNs (1776, 1777, and 1778 antisense ODNs) are targeted towards the stop codon region of the NPY Y5 mRNA.
3. Unmodified phosphodiester derivatives are injected in a combination of two sequences.
4. Each sequence is also applied as unmodified phosphodiester and phosphothioate-terminal-protected derivative.

### 3.2.2. Antisense ODNs Screening Protocols

*Basal food intake*

1. Rats are injected icv on three occasions with 50 µg antisense ODNs or 10 µL saline. Injections are performed at 10 AM, 1 PM and 4 PM.
2. Spontaneous food intake is recorded 22 h after the first icv injection.

*NPY-induced food intake*

The same NPY Y5 ODN sequences and derivatives are icv applied and tested for the effect on basal food intake. A saline pretreated group serves as a control for pNPY-induced feeding.

1. Rats are injected icv on six occasions over 2 days with 50 µg NPY Y5 antisense ODNs or 10 µL saline. Injections are performed at 10 AM, 1 PM, and 4 PM on day 1 and at 8 AM, 11 AM, and 2 PM on day 2.
2. NPY Y5 ODN pretreatment is followed by an icv injection of 300 pmol pNPY, performed 1 h after the last ODN injection.
3. Food intake is measured 1 h after pNPY application.

### 3.2.3. Screening Results

*Basal food intake*

The effects of three icv injections of 50 µg unmodified phosphodiester or phosphothioate-terminal-protected NPY Y5 mRNA antisense ODN sequences on 22-h food intake are compared with saline-treated rats and are depicted in **Fig. 2A**. Statistical analysis reveals that the unmodified phosphodiester antisense ODN sequences 1918 and 1776 and the phosphothioate-terminal protected derivatives 1917, 1918, and 1776 cause a significant decrease in food intake.

The reduction in food intake obtained with 1777 is probably due to toxicity, since rats treated with this antisense ODN showed dried blood secretion in the eyes and nose. These symptoms are even more pronounced with the phosphothioate-protected 1777 ODN. Therefore, no further experiments are performed with this antisense sequence (*see* **Notes 4.1.** and **4.4.**).

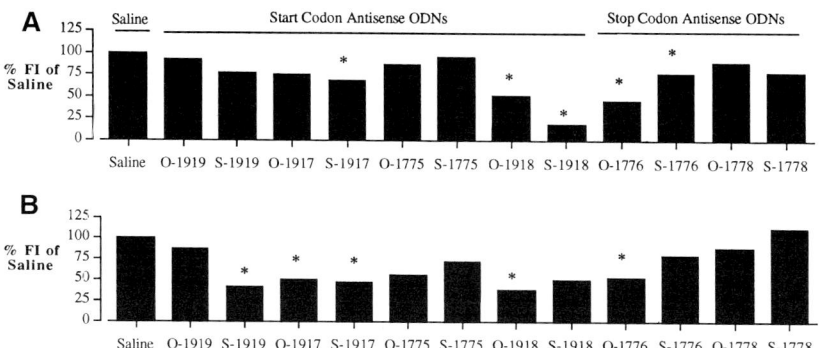

Fig. 2. Effect of different phosphodiester (O-) and phosphothioate-terminal-protected (S-) NPY Y5 antisense ODNs on basal and pNPY-induced food intake. (**A**) Rats were treated icv on three occasions with 50 μg of NPY Y5 antisense ODNs or 10 μL saline. Food intake was measured for a 22-h period. (**B**) Food intake was measured 1 h after pNPY (300 pmol) icv application in control rats, pretreated icv on six occasions over 2 days with 10 μL saline or 50 μg of NPY Y5 antisense ODNs. Results are expressed as % food intake (FI) of saline-treated rats. Each bar represents the mean of 7–16 observations. *, $p \leq 0.05$ vs saline. Statistical analyses were performed with the original data (absolute amount of food expressed in grams), and thus no statistical error bars are shown.

None of the combinations of two unmodified phosphodiester antisense ODNs demonstrated a stronger effect on spontaneous feeding when compared to the single compounds (data not shown; *see* **Note 4.1.**).

*NPY-induced food intake*

The effects of six consecutive icv injections of 50 μg unmodified phosphodiester or phosphothioate-terminal-protected antisense ODNs targeting the NPY Y5 mRNA on NPY-induced food intake are compared with saline-treated rats and are illustrated in **Fig. 2B**. The unmodified phosphodiester derivatives 1917, 1918, and 1776 as well as the phosphothioate-protected derivatives 1919 and 1917 significantly inhibited NPY-induced food intake. Similarly, in this experimental setup, combinations of two unmodified phosphodiester antisense ODNs did not exceed the effects of the single compounds (data not shown; *see* **Note 4.1.**).

### 3.2.4. Validation of the In Vivo Screening Model Using Appropriate ODN Control Sequences

*Basal food intake*

1. The effect of NPY Y5 antisense ODN treatment on basal food intake is studied with phosphothioate-terminal-protected antisense ODNs targeted to the start and stop codon regions of the NPY Y5 mRNA (1917 and 1776, respectively).

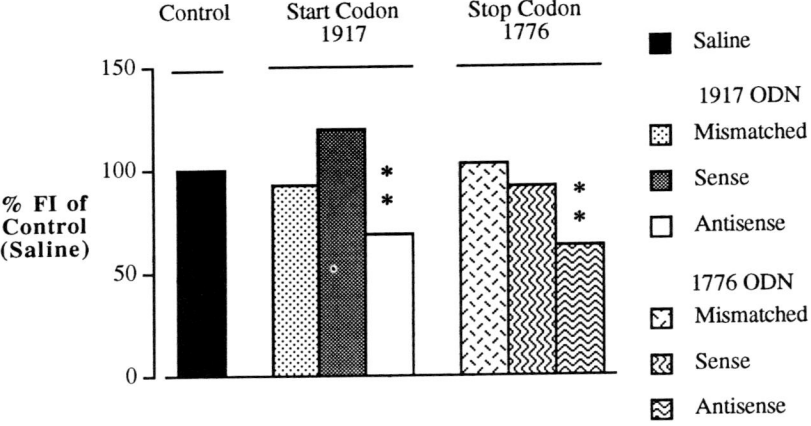

Fig. 3. Effect of two different phosphothioate-terminal-protected NPY Y5 antisense ODNs on basal food intake measured for a 22-h period. Rats were treated icv on three occasions with 10 μL saline, 50 μg of NPY Y5 mismatched, sense or antisense ODNs. Results are expressed as % food intake (FI) of saline-treated rats. Each bar represents the mean of 6–14 observations. **, $p \leq 0.01$ vs control. The absolute data have been published *(21)*.

2. Three icv injections of 50 μg antisense, mismatched, or sense ODNs are administered as described for the antisense ODN screening protocol (*see* **Subheading 3.2.2.**).

Both phosphothioate-terminal protected start and stop codon antisense sequences significantly reduced 22-h food intake when compared with saline-treated rats, mismatched, and sense treated rats (**Fig. 3**; *see* **Notes 4.2.** and **4.5.**

*NPY-induced food intake*

To study the effect of the phosphothioate-terminal-protected NPY Y5 antisense ODNs 1917 and 1778 and their mismatched and sense control sequences on NPY-induced food intake, the same ODN pretreatment is used as described for the antisense ODN screening protocol (*see* **Subheading 3.2.2.**). **Figure 4** illustrates the significant reduction of NPY-induced food intake after pretreatment with start and stop codon phosphothioate-terminal-protected antisense sequences when compared with their mismatched or sense control sequences or to saline (*see* **Notes 4.2.** and **4.5.**).

## 3.2.5. Use of Galanin as a Control for Specificity

1. Galanin-induced food intake in NPY Y5 antisense ODN pretreated rats is studied with the two NPY Y5 phosphothioate-terminal protected antisense ODNs 1917 and 1776.

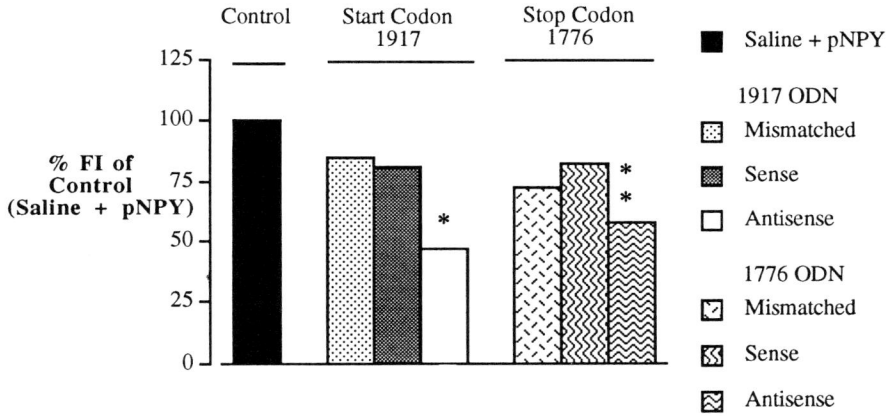

Fig. 4. Effect of two different phosphothioate-terminal-protected NPY Y5 antisense ODNs on pNPY-induced food intake. Food intake was measured 1 h after pNPY (300 pmol) icv application in control rats pretreated icv on six occasions over 2 days with 10 μL saline, 50 μg of NPY Y5 mismatched, sense, or antisense ODNs. Results are expressed as % food intake (FI) of control treated rats. Each bar represents the mean of 6–14 observations. **, $p \leq 0.01$; *, $p \leq 0.05$ vs control rats receiving pNPY. The absolute data have been published *(21)*.

2. For NPY Y5 antisense ODN pretreatment, the same protocol is used as described for the antisense ODN screening protocol (*see* **Subheading 3.2.2.**).
3. However, instead of pNPY, a dose of 2000 pmol galanin is icv applied to induce food intake.

No effect on galanin-induced food intake could be observed when rats were pretreated six times icv with 50 μg of 1917 or 1776 phosphothioate-terminal-protected antisense ODN when compared with saline treated rats (**Fig. 5**; *see* **Note 4.2.**).

### 3.3. Protocol for Ex Vivo NPY Y5 mRNA Quantification

#### 3.3.1. Isolation of mRNA from Rat Hypothalamic Tissue

1. Messenger RNA is isolated from the hypothalamic tissue using the QuickPrep Micro mRNA Purification Kit according to the manufacturer's instructions with a minor modification. Hypothalami in the present study are extracted in 1 mL lysis buffer, homogenized briefly with a Polytron (#PT-1200 C, Kinematica, Littau, Switzerland) tissue homogenizer and incubated at 56°C for 30 min.
2. The amount of mRNA is quantified by spectrophotometry and normalized to a concentration of 5 ng/μL.

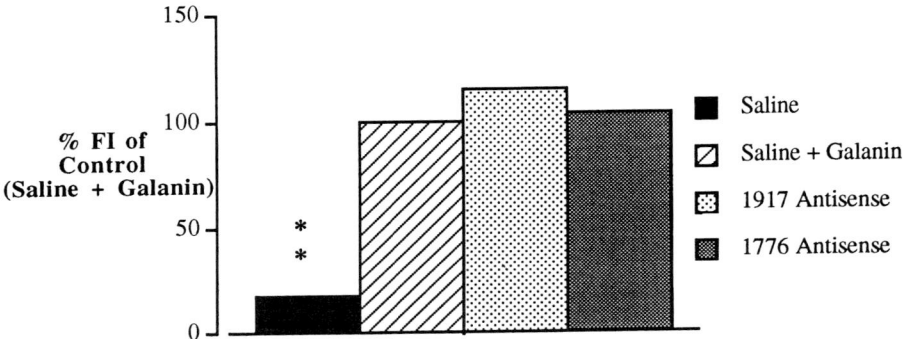

Fig. 5. Effect of two different phosphothioate-terminal-protected NPY Y5 antisense ODNs on galanin-induced food intake. Food intake was measured 1 h after galanin (2000 pmol) icv application in control rats pretreated icv on six occasions over 2 days with 10 µL saline (hatched bars), 50 µg 1917 (dotted bars), or 1776 NPY Y5 antisense ODNs (shaded bars). A saline pretreated and saline icv injected group served as control for galanin-induced food intake (filled bar). Results are expressed as % food intake (FI) of control rats. Each bar represents the mean of five to eight observations. $**, p \leq 0.01$ vs control rats receiving galanin. The absolute data have been published *(21)*.

### 3.3.2. Quantification of NPY Y5 mRNA by RT-PCR

*NPY Y5 mimic preparation*

1. To quantify RT-PCR products, heterologous standards/MIMICs for the rat Y5 are reverse transcribed and coamplified during the PCR reaction in a competitive fashion with the target cDNA *(20)*.
2. The internal standard (cRNA) for quantification of NPY-Y5 receptor mRNA is constructed in Bluescript BS-SK by insertion of a 30-bp poly(A) tail into *Not*I/ *Sac*I restriction sites in which a 2.0-kb fragment of the rat NPY-Y5 receptor cDNA (including a 0.7-kb 3' untranslated region) is subcloned.
3. A 150-bp DNA fragment from an unrelated gene is inserted into the *Bcl*I restriction site located within the PCR-amplified sequence.
4. NPY-Y5 receptor cRNA is prepared as a sense strand.
5. The RT-PCR product of the synthetic cRNA shows a size of 950 bp.

*RT-PCR*

1. First-strand cDNA synthesis is performed with 20 ng (4 µL) of isolated mRNA for each RT reaction.
2. 1 µL of decreasing amounts of cRNA (12.5–0.75 fg/µL) are added and heated for 5 min at 65°C and then chilled on ice.
3. Samples are incubated for 30 min at 42°C in the presence of 3.425 µL RT buffer for Y5 or β-actin mRNA quantification.

**Table 1**
**Effect of Two Different Phosphothioate-Terminal-Protected**
**NPY Y5 Antisense ODNs on Hypothalamic NPY Y5 mRNA Levels[a]**

| NPY Y5 | NPY Y5 mRNA (fg/mg protein) | β-Actin mRNA (ng/mg protein) | Ratio of NPY Y5 mRNA/β-actin mRNA |
|---|---|---|---|
| Saline | $3.15 \pm 0.19$ | $0.90 \pm 0.08$ | $3.62 \pm 0.41$ |
| 1917 antisense | $3.55 \pm 0.39$ | $1.16 \pm 0.18$ | $3.26 \pm 0.59$ |
| 1776 antisense | $2.03 \pm 0.30$ | $0.65 \pm 0.10$ | $3.18 \pm 0.11$ |

[a]Rats were pretreated icv on six occasions over 2 days with 10 μL saline, 50 μg NPY Y5 start codon (1917) or stop codon (1776) antisense ODNs. NPY Y5 mRNA was isolated from hypothalamic tissue and measured as described (*see* **Subheading 3.3.**). Results are expressed as fg/mg protein for NPY Y5 mRNA and as ng/mg protein for β-actin mRNA. The ratio of NPY Y5 mRNA/β-actin mRNA is a parameter without units. Each value represents the mean ± SEM of five observations.

4. The reaction is stopped by heating the samples for 5 min at 95°C and chilling the tubes on ice.
5. PCR is carried out with Taq DNA polymerase.
6. PCR buffer, 41.25 μL for Y5 and β-actin mRNA quantification are added to each tube.
7. 45 PCR cycles consisting of denaturation at 94°C (30 s), annealing at 63°C (20 s), and extension at 72°C (30 s) are performed.
8. PCR is completed by a final extension step of 5 min at 72°C.
9. The amplification products originating from the NPY-Y5 mRNA or from the internal standard are separated by electrophoresis on 1.5% agarose gels and stained with ethidium bromide.
10. Fluorescence is recorded using the Cybertech CS system.
11. The ratio of the signal generated by the amplification products is quantified, and the log of the ratios are plotted as a function of the initial amount of NPY-Y5 cRNA.
12. The amount of NPY-Y5 mRNA present is estimated by linear regression (cRNA amount for which the log of the ratio is equal to zero), and the values are normalized to β-actin mRNA.

*NPY Y5 mRNA measurement*

To investigate whether NPY Y5 antisense treatment could affect the quantity of Y5 receptor mRNA in vivo, rats are pretreated with two phosphothioate-terminal-protected antisense ODNs (1917 and 1776) as described for the antisense ODN screening protocol (*see* **Subheading 3.2.2.**). NPY Y5 mRNA levels tend to be only slightly decreased, when six icv injections of 50 μg of 1917 and 1776 phosphothioate-terminal-protected antisense ODNs are administered (**Table 1**).

## 3.4. In Vitro Studies: Protocols
## for Using a Surrogate Cell Model

### 3.4.1. Cell culture

1. The effect of ODN treatment on Y5 receptor synthesis is studied in human embryonic kidney cells (HEK-293) and in a mouse fibroblast cell line (LMTK), stably transfected with the rat NPY Y5 receptor *(5)*.
2. Cells are maintained in 175-cm² flasks at 37°C in cell culture medium.
3. When cells are confluent (about every 3–4 days), they are passaged after enzymatic disaggregation with trypsin-EDTA at a ratio of 1:20.
4. The medium is removed by aspiration, and the cells are washed with 20 mL PBS and incubated with 5 mL trypsin-EDTA for 3 min at 37°C.
5. The reaction is stopped by addition of 15 mL DMEM (at a total volume of 20 mL) and 1.5 mL of this solution is plated at a density of approximately 40,000 cells/cm² in a culture flask.
6. The medium is changed every 2 days until cells are confluent again.
7. Cell numbers are obtained by counting aliquots of the cell suspensions after trypsinization in isotonic solution with a Coulter counter (Coulter, England).

### 3.4.2. Radioligand Binding Assay in Cell Culture

1. NPY Y5 receptors are quantified using a radioligand binding assay.
2. [$^{125}$I-Pro34]hPYY is used as radioligand, and nonspecific binding is determined using an excess (1 µ$M$) of unlabeled [Pro34]hPYY.
3. Cells are preincubated for 30 min with 250 µL/well of incubation medium.
4. 10 µ$M$ unlabeled [Pro34]hPYY for nonspecific binding or HEPES for total binding and 50 µL [$^{125}$I-Pro34]hPYY (final concentration 25 pM for HEK-293-cells, 150 p$M$ for LMTK cells) are added to each well.
5. 2 h of incubation at room temperature.
6. Incubation mixture is removed and cells are washed three times with 500 µL ice-cold PBS.
7. Cells are dissolved in 500 µL 4 $M$ NaOH for 12 h.
8. Aliquots of 400 µL are taken for counting (LKB-Wallac 1277 GammaMaster, Turku, Finland) and for subsequent protein determination.

### 3.4.3. Studies with NPY Y5 ODNs in HEK-293 and LMTK Cells

*Cell culture studies in HEK-293 cells*

1. Incubation plates (Costar®, Corning Costar, Cambridge, MA) are pretreated for 2 h at room temperature with 200 µL coating reagent.
2. Wells are washed twice with sterile PBS.
3. 300 µL of a HEK-293 rY5 cell suspension (200,000 cells/mL DMEM) is plated in 24-well Costar plates and incubated for 24 h at 37°C under normal culture conditions.
4. Cells are washed twice with PBS.
5. Opti-MEM is added in the absence or presence of ODNs (1 and 10 µ$M$, with 0.2 and 2% transfection reagent, respectively).

**Table 2**
**Effects of 24 Hours of Treatment with Two Different NPY Y5 Phosphothioate-Terminal-Protected Antisense ODNs on $B_{max}$ (fmol/mg protein) of NPY Y5 Receptors in HEK-293 Cells[a]**

| NPY Y5-S-ODNs | $B_{max}$ (fmol/mg protein) |
|---|---|
| 1917-S-ODNs (1 μ*M*) | |
| Antisense | 9.6 ± 1.1* |
| Mismatched | 11.6 ± 0.9 |
| 1917-S-ODNs (10 μ*M*) | |
| Antisense | 7.6 ± 0.3* |
| Mismatched | 11.1 ± 1.2 |
| 1776-S-ODNs (1 μ*M*) | |
| Antisense | 8.0 ± 0.6 |
| Mismatched | 8.6 ± 0.8 |
| 1776-S-ODNs (10 μ*M*) | |
| Antisense | 6.4 ± 0.7* |
| Mismatched | 8.5 ± 0.7 |

[a]Cells were treated with 1 or 10 μM of antisense ODNs targeting the start condon (1917) or the stop condon (1776) of the NPY Y5 mRNA or mismatched ODNs. $B_{max}$ (fmol/mg protein) of NPY Y5 receptors were determined after a 24-h incubation with NPY Y5 ODNs as described in **Subheading 3.4.2.** Each value represents the mean ± SEM of four to six observations.
*, p ≤ 0.05 vs mismatched ODN.

6. Incubation is continued for a further 24 h at 37°C, after which the cells are washed once with 500 μL DMEM.

**Table 2** shows a significant reduction of the $B_{max}$ in HEK-293 cells after treatment with 1 and 10 μ*M* of the phosphothioate-terminal-protected antisense ODN 1917 when compared with the mismatched ODN-treated group. When cells were treated with the same derivative of 1776 ODN, $B_{max}$ was slightly reduced after 1 μ*M*, but significantly after 10 μ*M* of antisense ODN treatment (*see* **Notes 4.3.** and **4.5.**).

*Cell culture studies in LMTK cells*

1. Studies in LMTK cells are performed using DOTAP as a liposomal transfection reagent, and preparation of the ODNs is performed according to the manufacturer's instructions.

**Table 3**
**Effects of Five Days of Treatment with Phosphothioate-Terminal-Protected Start (1917 and 1918)**
**and Stop Codon (1776) NPY Y5 Antisense ODNs**
**on $B_{max}$ (fmol/mg protein) of NPY Y5 Receptors in LMTK Cells[a]**

| NPY Y5-S-ODNs | $B_{max}$ (fmol/mg protein) (3 µ$M$) | $B_{max}$ (fmol/mg protein) (10 µ$M$) |
|---|---|---|
| 1917-S-ODNs | | |
| Antisense | 6.0 ± 0.8 | 7.5 ± 1.5 |
| Sense | 7.5 ± 0.7 | 7.2 ± 1.2 |
| Mismatched | 7.2 ± 0.4 | 6.2 ± 1.1 |
| DOTAP | 6.2 ± 0.7 | 6.6 ± 1.1 |
| 1918-S-ODNs | | |
| Antisense | 9.5 ± 0.9 | |
| Sense | 8.5 ± 0.7 | |
| Mismatched | 9.1 ± 0.6 | |
| DOTAP | 9.0 ± 0.6 | |
| 1776-S-ODNs | | |
| Antisense | 7.4 ± 1.0 | 4.8 ± 2.0 |
| Sense | 8.2 ± 0.7 | 3.4 ± 1.6 |
| Mismatched | 8.7 ± 0.8 | 3.7 ± 2.1 |
| DOTAP | 6.1 ± 1.9 | 4.6 ± 0.9 |

[a]Concentrations of 3 (1917, 1918, and 1776) or 10 µ$M$ (1917 and 1776) of NPY Y5 antisense, sense, or mismatched ODNs or DOTAP were cumulatively added to cells for five consecutive days. $B_{max}$ (fmol/mg protein) of NPY Y5 receptors was determined on day 6 as described in **Subheading 3.4.2.** Each value represents the mean ± SEM of 8–10 observations.

2. 300 µL of an LMTK rY5 cell suspension (50,000 cells/mL DMEM) is plated in 24-well incubation plates (Biocoat, Becton Dickinson, Bedford, MA), coated with synthetic poly-D-lysin.
3. A 24-h incubation at 37°C under normal culture conditions is followed by two washes of the cells with PBS.
4. Opti-MEM is added in the absence or presence of ODNs (3 or 10 µ$M$ with 2% DOTAP) to each well from day 1 to 6 (300 µL on the first day and 60 µL on the following 5 days).
5. Incubation is performed at 37°C for the duration of treatment.

NPY Y5 binding sites of LMTK cells were differentially affected after treatment with 3 or 10 µ$M$ 1917, 1918, or 1776 phosphothioate-terminal-protected ODNs (**Table 3**). In addition, treatment with the start codon sequences showed divergent effects (**Table 3**). The $B_{max}$ of LMTK cells tended to be decreased after treatment with the antisense ODN sequence 1917, but increased when the antisense sequence 1918 was added to the cells. The 1776 antisense ODN only tended to reduce $B_{max}$ when compared with control ODN-treated cells. A trend toward an increase in $B_{max}$ could be observed for cells treated with DOTAP. No effect on $B_{max}$ was obtained when 10 μ$M$ of 1917 antisense ODN were applied. When cells were treated with 10 μ$M$ of 1776 antisense ODN, a small but not significant increase in $B_{max}$ could be observed compared with control ODN, but similar levels were seen for DOTAP treated cells (*see* **Subheadings 4.3.** and **4.5.**).

## 4. Notes

### 4.1. Antisense ODN Screening Strategy

1. In a first step we identified active antisense sequences using an in vivo screening model: Among 12 different NPY Y5 start and stop codon antisense ODNs, only 3 showed significant reductions in spontaneous and NPY-induced food intake (**Fig. 2**). The only sequence that induced overt toxic symptoms was the phosphothioate-terminal-protected stop codon sequence 1777: rats showed blood-stained secretions in eyes and nose. These data support the observations from other investigators, suggesting that among different antisense ODN sequences to a defined mRNA, only a few sequences target the crucial site on the gene, interfering with gene expression whereas others are inactive or might even induce toxic effects *(16)* (*see* **Subheading 3.2.3.**).

2. The hypothesis that a combination of two antisense sequences would lead to a potentiated effect was proved false since all combinations tested failed to exceed the effects of a single antisense sequence (data not shown; *see* **Subheading 3.2.3.**).

### 4.2. Validation of the In Vivo Screening Strategy

3. To rule out nonspecific effects and moderate to light toxicity of the active NPY Y5 antisense ODNs, a comparison with the respective control sequences was needed. The selection criteria used was based on the assumption that active NPY Y5 antisense ODNs should reduce spontaneous, and NPY-induced food intake to a similar extent. Second, we postulated that both derivatives should be active in at least one model. For subsequent in vivo experiments, the phosphothioate-protected derivatives were selected because they are more stable in vivo than the unmodified derivatives.

4. In the subsequent in vivo experiments, the start and stop codon antisense ODNs significantly reduced basal food intake as well as NPY-induced food intake. No effect was observed with the respective control sequences *(21)* (**Figs. 3** and **4**; *see* **Subheading 3.2.4.**).

5. The specificity of this effect was supported since neither the start nor the stop codon NPY Y5 antisense ODNs affected galanin-induced food intake *(21)* (**Fig. 5**; *see* **Subheading 3.2.5.**).

## 4.3. Investigation of the Mechanism of Antisense Action

6. The mechanism of antisense action could not be fully documented since the lack of a specific peptidic ligand precluded the ex vitro determination of NPY Y5 binding sites.

7. RT-PCR on rat hypothalamic tissue after NPY Y5 antisense treatment showed only a trend for reduced NPY Y5 mRNA expression levels (**Table 1**). Since only one set of primers was used for RT-PCR, changes in NPY Y5 mRNA may have been missed. The primers chosen could in principle amplify partially degraded mRNA. More extensive RT-PCR studies using primers for the regions targeted by the antisense ODN would have increased the quality of these studies (see **Subheadings 2.1.** and **2.3.**).

8. Additional ex vivo studies using different techniques than RT-PCR could provide information about an involvement of RNase H. For example, a ribonuclease protection assay would allow quantitative determination of the target mRNA. Quantitative as well as local changes of NPY Y5 mRNA could be observed using *in situ* hybridization.

9. Our findings suggest that other mechanisms than RNase H-mediated cleavage of the target mRNA might be responsible for the observed in vivo antisense effect on food intake. Hybrid-arrested translation of the protein could be the underlying mechanism of action because of the following observation: in HEK-293 cells, a cell line expressing low levels of Y5 receptors, 24 h-treatment with 1 and 10 $\mu M$ of a NPY Y5 start codon sequence and 10 $\mu M$ of a NPY Y5 stop codon antisense ODN significantly reduced NPY Y5 receptor densities (**Table 2**). Unfortunately, studies in LMTK cells revealed inconsistent data with different concentrations of the start and stop codon sequences despite the use of a transfection reagent (**Table 3**). Apparently, in vitro activity of phosphothioate-protected antisense ODNs does not necessarily depend on the use of a transfection reagent but is strongly influenced by the cell system (*see* **Subheading 3.4.**).

## 4.4. Potential and Pitfalls of Antisense ODNs

10. The suitability of our antisense approach was demonstrated in studies using the antisense sequences described above to characterize the role of the NPY Y5 receptor in NPY-mediated feeding. The NPY Y5 antisense ODNs inhibited spontaneous, food deprivation, and NPY-induced feeding in rats *(21)*. These data have since been confirmed by another group, using an unmodified 20-bp start codon antisense sequence and its scrambled control *(22)*. These findings are further supported by recent studies using the highly selective nonpeptidic NPY Y5 antagonist CGP 71683A, which had comparable effects on spontaneous, food deprivation, and NPY-induced feeding in rats *(23)*.

11. Another option for studying the role of the NPY Y5 receptor in NPY-induced feeding is the NPY Y5 knockout mouse. However, possible physiological compensatory mechanisms occurring in knockout animals challenge the

investigation of new genes. This may also explain the observation of Marsh et al. *(24)*, who reported that NPY Y5 receptor knockout mice feed and grow normally. Furthermore, no effect could be observed on fasting-induced food intake in the absence of the NPY Y5 receptor *(24)*, whereas pretreatment with an NPY Y5 start codon antisense ODN significantly decreased fasting-induced food intake in rats *(21)*. The reduction of NPY-induced feeding in NPY Y5 knockout mice suggests that the NPY Y5 receptor subtype mediates at least some of the effects of centrally applied NPY on feeding. These data are consistent with our finding that pretreatment with start and stop codon NPY Y5 antisense ODNs significantly reduced NPY-induced food intake *(21)*. These studies indicate that the antisense approach, often referred to as the "knock-down" technique, may be a less dramatic change for the system studied. Another advantage would be that the gene knock-down can be constantly or intermittently performed at selected time points during the lifetime of the animal.

12. The encouraging data from our antisense ODN experiments contrast with numerous reports about the lack of efficacy or specificity of antisense ODNs. For example, Dryden et al. *(25)* showed in a recent study that unmodified phosphodiester ODNs designed to inhibit NPY gene expression are not detectable in the hypothalamus after 7 days of administration into the third ventricle and had no effects on food intake. Phosphothioates, on the other hand, were widely distributed, but severely toxic. Our observations with one stop codon phosphothioate antisense sequence causing bleeding in the nose and the eyes are consistent with this report (*see* **Subheading 3.2.3.**). Toxicity caused by ODN degradation products or by phosphothioate modification clearly limited our studies, but the choice of a phosphothioate-terminal protection over a full backbone protection decreased the risk of phosphothioate-induced toxicity. Nonetheless, at higher doses toxic symptoms were also observed in our studies and precluded the completion of a full dose-response curve or extended treatment. Such restrictions limited the investigation of the maximum effect of NPY Y5 receptor downregulation.

## 4.5. Improvement of the NPY Y5 Antisense Approach

13. Possible interpretations of the NPY Y5 antisense approach are limited for several reasons. First, food intake is a critical parameter, since sick animals decrease their food intake. Possible toxic effects of the sequences tested could have led to false-positive results. However, careful observation of the treated animals during the screening procedure and the use of the appropriate control sequences minimized this risk (*see* **Subheading 3.2.4.**). The use of a highly selective nonpeptidic NPY Y5 antagonist such as CGP 71683A in NPY Y5 antisense ODN pretreated animals could have been helpful. A decreased activity of the antagonist in such rats would further support a specific effect of these ODNs sequences on NPY Y5 receptors. Conversely, studies with the same sequences in NPY Y5 knockout mice would serve the same purpose.

Second, the lack of a specific NPY Y5 agonist precluded the final proof of decreased NPY Y5 binding sites after antisense treatment. Receptor binding studies in two different cell lines were performed to compensate for this fact (*see* **Subheading 3.4.**). The divergent results obtained in these experiments further supported the evidence that the final proof of decreased NPY Y5 binding sites can only be obtained with ex vivo studies.

14. The weakness of the approach described in this chapter was that we dealt with two unknown parameters at the beginning of the experiments: the functional role of the NPY Y5 receptor subtype as well as active NPY Y5 antisense sequences (*see* **Subheading 3.1.**). Despite this fact, we based the selection criteria for active antisense sequences on the hypothesis of Gerald et al. *(5)* that the NPY Y5 receptor subtype was involved in the mechanism of NPY-induced feeding. The fact that among the several antisense sequences tested, only three could be selected for further studies supports the notion that even when hybridization to the target mRNA is predicted, a specific in vivo, ex vivo, or in vitro effect is not necessarily obtained (*see* **Subheading 3.2.2.**).

Taken together, effects on basal and NPY-induced food intake were used to identify active phosphothioate-terminal-protected NPY Y5 antisense ODNs. These ODN sequences were used in subsequent experiments performed to characterize the NPY Y5 receptor. Receptor binding studies in different cell lines and ex vivo determination of receptor mRNA expression suggest that hybrid-arrested translation at the protein level might be the underlying mechanism of antisense action. The data obtained with antisense ODNs on basal and NPY-induced food intake in vivo show the suitability of antisense ODNs as pharmacological tools in testing biological hypotheses. Nevertheless, the ex vivo and in vitro data indicate the difficulty of demonstrating a specific effect. In future, these tools may be of interest to investigate the physiological role of many G-protein-coupled receptors found in the hypothalamus as long as their possible limitations are kept in mind.

## References

1. Blomqvist, A. G. and Herzog, H. (1997) Y-receptor subtypes: how many more? *Trends Neurosci.* **20,** 294–298.
2. Bard, J. A., Walker, M. W., Branchek, T. A., and Weinshank, R. L. (1995) Cloning and functional expression of a human Y4 subtype for pancreatic polypeptide, neuropeptide Y and peptide YY. *J. Biol. Chem.* **270,** 26,762–26,765.
3. Gehlert, D. R., Beavers, L. S., Johnson. D., Gackenheimer, S. L., Schober, D. A. and Gadski, R. A. (1996) Expression cloning of a human brain neuropeptide Y Y2 receptor. *Mol. Pharmacol.* **49,** 224–228.
4. Gerald, C., Walker, M. W., Vaysse, P. J., Branchek, T. A., and Weinshank, R. L. (1995) Expression cloning and pharmacological characterization of a human hippocampal neuropeptide Y/peptide YY Y2 receptor subtype. *J. Biol. Chem.* **270,** 26,758–26,761.

5. Gerald, C., Walker, M. W., Criscione, L., Gustafson, E. L., Batzl Hartmann, C., Smith, K. E., et al. (1996) A receptor subtype involved in neuropeptide-Y-induced food intake. *Nature* **382,** 168–171.
6. Gregor P., M. L. Millham, Y. Feng, L. B. DeCarr, M. L. McCaleb and L. J. Cornfield (1996) Cloning and characterization of a novel receptor to pancreatic polypeptide, a member of the neuropeptide Y receptor family. *FEBS Lett.* **381,** 58–62.
7. Larhammar, D., Blomqvist, A. G., Yee, F., Jazin, E., Yoo, H., and Wahlestedt, C. (1992) Cloning and functional expression of a human neuropeptide Y/peptide YY receptor of the Y1 type. *J. Biol. Chem.* **267,** 10,935–10,938.
8. Lundell, I., Blomqvist, A. G., Berglund, M. M., Schober, D. A., Johnson, D., Statnick, M. A., et al. (1995) Cloning of human receptor of the NPY receptor family with high affinity for pancreatic polypeptide and peptide YY. *J. Biol. Chem.* **270,** 29,123–29,128.
9. Weinberg, D. H., Sirinathsinghji, D. J., Tan, C. P., Shiao, L. L., Morin, N., Rigby, M. R., et al. (1996) Cloning and expression of a novel neuropeptide Y receptor. *J. Biol. Chem.* **271,** 16,435–16,438.
10. Hu, Y., Bloomquist, B. T., Cornfield, L. J., DeCarr, L. B., Flores Riveros, J. R., Friedman, L., et al. (1996) Identification of a novel hypothalamic neuropeptide Y receptor associated with feeding behavior. *J. Biol. Chem.* **271,** 26,315–26,319.
11. Stanley, B. G., Magdalin, W., Seirafi, A., Nguyen, M. M. and Leibowitz, S. F. (1992) Evidence for neuropeptide Y mediation of eating produced by food deprivation and for a variant of the Y1 receptor mediating this peptide's effect. *Peptides* **13,** 581–587.
12. Wyss P., Stricker-Krongrad, A., Brunner, L., Miller, J., Crossthwaite, A., Whitebread, S., et al. (1998) The pharmacology of neuropeptide Y (NPY) receptor mediated feeding in rats characterizes better Y5 than Y1, but not Y2 or Y4 subtypes. *Regul. Pept.* 75–**76,** 363–371.
13. Gold, L. H. (1996) Integration of molecular biological techniques and behavioural pharmacology. *Behav. Pharmacol.* **7,** 589–615.
14. Whitesell, L., Geselowitz, D., Chavany, C., Fahmy, B., Walbridge, S., Alger, J. R., et al. (1993) Stability, clearance, and disposition of intraventricularly administered oligodeoxynucleotides: implications for therapeutic application within the central nervous system. *Proc. Natl. Acad. Sci. USA* **90,** 4665–4669.
15. Wahlestedt, C. (1994) Antisense oligonucleotide strategies in neuropharmacology. *Trends Pharmacol. Sci.* **15,** 42–45.
16. Monia, B. P., Johnston, J. F., Geiger, T., Muller, M., and Fabbro, D. (1996) Antitumor activity of a phosphorothioate antisense oligodeoxynucleotide targeted against C-raf kinase. *Nature Med.* **2,** 668–675.
17. Helene, C. and Toulme, J. J. (1990) Specific regulation of gene expression by antisense, sense and antigene nucleic acids. *Biochem. Biophys. Acta* **1049,** 99–125.
18. Morris, M. and Lucion, A. B. (1995) Antisense oligonucleotides in the study of neuroendocrine systems. *J. Neuroendocrinol.* **7,** 493–500.
19. Paxinos G. and Watson, C. (1986) *The Rat Brain Stereotaxic Coordinates.* Academic, London.

20. Lesniak W., Schaefer, C., Grueninger, S., and Chiesi, M. (1995) Effect of alpha adrenergic stimulation and carnitine palmitoyl transferase I inhibition on hypertrophying adult rat cardiomyocytes in culture. *Mol. Cell. Biochem.* **142,** 25–34.
21. Schaffhauser A. O., Sricker-Krongrad, A., Brunner, L., Cumin, F., Gerald, G., Whitebread, S., et al. (1997) Inhibition of food intake by neuropeptide Y5 antisense oligodeoxynucleotides. *Diabetes* **46,** 1792–1798.
22. Tang-Christensen M., Kristensen, P., Stidsen, C. E., and Brand, C. L. (1998) Central administration of Y5 receptor antisense decreases spontaneous food intake and attenuates feeding response to exogenous neuropeptide Y. *J. Neuroendocrinol.* **159,** 307–312.
23. Criscione L., Rigollier, P., Batzl-Hartmann, C., Rüeger, H., Stricker-Krongrad, A., Wyss, P., et al. (1998) Food intake in free-feeding and energy-deprived lean rats is mediated by the neuropeptide Y5 receptor. *J. Clin. Invest.* **102,** 2136–2145.
24. Marsh D. J., Hollopeter, G., Kafer, K. E., and Palmiter, R. D. (1998) Role of the Y5 neuropeptide Y receptor in feeding and obesity. *Nature Med.* **4,** 718–721.
25. Dryden S., Pickavance, L., Tidd, D., and Williams, G. (1998) The lack of specificity of neuropeptide Y (NPY) antisense administered intracerebroventricularly in inhibiting food intake and NPY gene expression in the rat hypothalamus. *J. Endocrinol.* **157,** 169–175.

# 12

## NPY Antisense Oligodeoxynucleotides to Study the Actions of NPY

### Pushpa S. Kalra and Satya P. Kalra

## 1. Introduction

Antisense oligodeoxynucleotides (ODN) that disrupt gene function, leading eventually to a reduction in levels of the encoded protein, are powerful molecular tools for studying the functions of neuropeptides and their receptors in the brain. Their potential usefulness is even greater for the ever increasing number of newly discovered peptide and receptor genes expressed in the central nervous system for which pharmacological blocking agents or selective ligands may not be available. Generally, antisense ODNs are short DNA sequences complementary to a specific mRNA that has the same sequence as the sense DNA strand. Although the mechanisms of antisense action are not fully understood, it prevents translation of the mRNA into protein by action at any one of several sites in the sequence of events from DNA to protein synthesis. This inhibition may occur (1) by binding to DNA to form a triple-helical structure; (2) by simply hybridizing to the target RNA and thus preventing its interaction with ribosomes, polymerases, and so forth; and (3) by binding to mRNA to provide a substrate for enzymes such as RNase-H that cleave and degrade the mRNA and prevent protein expression. Additionally, there is evidence that antisense ODNs may bind directly to the target proteins, resulting in its inhibition. The latter effect (referred to as aptomer binding) occurs in a non-antisense manner, however, inhibition of a specific protein may be useful from a therapeutic point of view. Whatever may be the mechanism of antisense ODN action, it is a powerful alternative approach to assess protein function, especially in the absence of pharmacological blockers.

Factors that govern choice of an antisense sequence include specificity, ease of cellular uptake, intracellular stability, and degradation. ODNs are readily

From: *Methods in Molecular Biology, vol. 153: Neuropeptide Y Protocols*
Edited by: A. Balasubramaniam © Humana Press Inc., Totowa, NJ

internalized into cells by endocytosis; however, because uptake is dependent on cell type and size of the sequence, these factors must be considered. Additionally, for in vivo studies, solubility, toxicity, and mode of administration of the ODN are vital considerations.

NPY is widely distributed in the brain and brainstem and has a multitude of neuroendocrine and behavioral actions. Within the hypothalamus it is the most abundant neuropeptide, produced primarily by neurons in the arcuate (ARC) nucleus and transported to various hypothalamic sites including the preoptic area, paraventricular nuclei (PVN), and the perifornicle area (PFH) which also receive NPY input from catecholamine-producing neurons in the brainstem *(1)*. NPY is intimately involved in regulating the release of pituitary luteinizing hormone (LH) by modulating the secretion of LH-releasing hormone (LHRH) from the hypothalamus *(2)*. Additionally, NPY has powerful orexigenic effects by action at the PVN and PFH areas *(3,4)*. Numerous studies have documented that NPY gene transcription in the ARC region and NPY levels in various hypothalamic sites fluctuate in accordance with the need for its action. NPY mRNA is upregulated in the ARC in conditions of high energy demand such as lactation, experimentally induced diabetes, fasting, and food restriction *(5–11)*. It is also upregulated during the period before the preovulatory surge release of pituitary LH to stimulate hypothalamic LHRH hypersecretion *(12)*. Thus, NPY antisense ODNs, which would reduce availability of the peptide by translational arrest, are well tailored for determining the physiological role of hypothalamic NPY in stimulating appetite as well as LH secretion.

## 2. Materials
### 2.1. ODNs

Antisense oligos can be readily and inexpensively synthesized by commercial oligodeoxynucleotide synthesizing laboratories according to the user's specifications. The cost is based on the number of nucleotides and modifications to the molecule. NPY ODNs must be delivered directly into the brain; thus it is important that the ODN be desalted and purified by high-performance liquid chromatography (HPLC) or ion exchange chromatography to remove all impurities, especially organic matter and solvents that may be toxic to neurons. This purification is performed by the company synthesizing the ODN. ODNs are stable and can be stored indefinitely at $-20°C$ in the lyophilized form. They are readily soluble in aqueous solutions and are dissolved in sterile saline both for intracerebroventricular administration and for microinjection into specific hypothalamic sites. The dissolved ODNs may be stored for several days at $4°C$; however, the best strategy is to aliquot the ODN before lyophilization so that it may be dissolved as needed in small batches.

## 2.2. Animal Preparation

For NPY antisense ODN delivery, animals are preimplanted with permanent guide cannulae into the third or lateral cerebroventricle. Guide cannulae (22 gauge, #C313G), 28-gage inner stylets (#C313I), and injection cannulae (#C313DC) are available from Plastics One (Roanoke, VA). The guide cannulae are threaded so that the inner stylet screws into the guide cannula to prevent cerebrospinal fluid (CSF) efflux and is replaced by the inner injection cannula for injection of ODNs. This setup allows an animal to be injected multiple times. The injection cannula is attached to a 50 µL Hamilton microsyringe by a 12-inch length of PE20 tubing. Small 1/8-inch machine screws (#R-MX-080-2) for anchoring the cannulae to the skull are purchased from Small Parts, (Miami Lakes, FL). Surgery is performed in a Kopf (Tujunga, CA) stereotaxic apparatus for small rodents.

## 3. Methods

### 3.1. Preparation of ODNs

#### 3.1.1. Design of NPY antisense ODNs

Theoretically, an antisense ODN of 17 nucleotides should be unique in the genome conferring specificity. Longer sequences would impart greater specificity but at the expense of cell penetration and are also subject to mismatches. Thus, conventionally, antisense ODNs range from 15 to 20 bases. This length is also optimal for neuronal uptake *(13)*. Although not essential, it is generally recommended that the antisense regions selected encompass the AUG start codon and extend to adjoining bases. For the rat NPY sequence (GENBANK Accession #M15880; *14*), the start codon for the signal peptide is at base 69, and several of the published antisense sequences are targeted to this region. Segments known to be toxic, such as G-tetramers or CpG are avoided. The selected sequence is analyzed in the antisense direction to avoid any self-complementarity, hairpin loops, and homology to known gene sequences using any appropriate computer programs (BLAST, CLUSTAL, PRIMER, and so forth). (*See* **Note 1.**)

#### 3.1.2. Control ODNs

Although in theory antisense technology should be highly specific in disrupting the targeted gene, in practice ODNs can cause toxicity and produce nonspecific effects (*see* **Note 2**). It is essential, therefore, to address the toxic and nonspecific effects rigorously by using several controls. The initial recommendation to use a sense control sequence of the same length as the antisense sequence has been largely rejected due to numerous nonspecific effects of the sense strand that complicate interpretation of antisense results. A better alternative is to use scrambled and mismatched sequences with the same length and base composition as the antisense sequence and a saline control. As few as two mismatched bases in the ODN sequence would prevent hybridization with mRNA and serve as an appropriate control. The scrambled sequences also have the advantage of keeping the same GC composition. As with the antisense ODN, the control sequences should be screened for homology to known genes as well as for self-hybridization.

### 3.1.3. Modification of ODN Constructs

Although relatively stable in the CSF following a bolus injection, the natural phosphodiester backbone of the sequence is rapidly degraded within minutes by neuronal α-exonucleases *(13)* (*see* **Note 3**). Chemical modification, by replacement of the phosphodiester linkage with phosphorothioate, i.e., replacing a nonbinding O atom at the 2' position by S, has been widely used to retard degradation. Phosphorothioated ODNs may bind directly to RNase H and thus inhibit its catalytic action and increase the half-life of the ODN in vivo *(15–18)*. There is no difference in initial uptake of the phosphodiester and phosphorothioate forms by the brain. Dryden et al. *(19)* directly compared the neuronal uptake and stability of a fluorescein-labeled NPY antisense sequence with three different structural modifications. Even after 7 days of constant infusion into the third ventricle via Alzet® osmotic minipumps, the unmodified phosphodiester was not detectable in the hypothalamus, thus confirming rapid degradation in vivo. The phosphorothioated ODN was much more stable and was widely distributed in the hypothalamus. This modification, however, produced adverse physical effects; the treated rats were listless and hunched and did not display grooming behavior. With a 3'-propyl substitution, the antisense ODN displayed good cellular uptake in the hypothalamus and stability without producing toxic effects *(19)*. In view of these counterbalanced requirements of enhanced stability in vivo without producing toxicity, a viable alternative is to phosphorothioate 1–2 bases at the termini (capping). Since antisense degradation starts at the 3'-end, protection of bases at this end alone can reduce degradation *(20)*. Both the phosphodiester and phosphorothioate forms of NPY antisense ODNs have been used to block its behavioral and neuroendocrine effects (*see* **Subheading 3.3.**).

### 3.2. Cannulations and Delivery

### 3.2.1. Intracerebroventricular Cannulation

The localized concentration of NPY-producing neurons in the ARC is amenable to blockade by intracerebroventricularly (icv) administered ODNs (*see* **Note 4**) . NPY ODNs have been injected into both the lateral and third cerebroventricle with no apparent difference in efficacy due to route of administration. Permanent guide cannulae for icv injection are stereotaxically implanted 5–7 days before the experiment, and the rats are allowed to recover from the surgical stress.

### 3.2.1.1. LATERAL VENTRICLE CANNULATIONS

Position the nose bar of the stereotaxic apparatus at 0.3 mm below 0. Anesthetize rats with phenobarbital or ketamine/xylazine, and fit into the stereotaxic frame. Drill a small hole into the skull at 8.2 mm anterior to the interaural line and 1.5 mm lateral to the midline suture. Lower the cannula 3.5 mm below the dura, and anchor in place with dental cement spread on the skull to encompass the cannula and three machine screws screwed into the skull around the cannula. The stylet is inserted to prevent CSF efflux, and the rat is allowed to recover in a warm environment.

### 3.2.1.2. THIRD VENTRICLE CANNULATION

Position the nose bar at +0.5 mm, and drill a hole at midline in the skull at 6.4 mm anterior to the interaural line. Lower the cannula 8.0 mm from the dura. Bleeding caused by rupture of the sagittal sinus is controlled by a small piece of Gelfoam placed around the cannula. The cannula is then secured in place as above, and the stylet is inserted.

## 3.2.2. Delivery of NPY Antisense ODN

1. The ODNs are dissolved in sterile saline or artificial CSF and injected in a volume of 3 µL (third ventricle) or 5 µL (lateral ventricle) via the preimplanted cannula in unanesthetized rats (*see* **Note 4**).
2. The ODNs may also be microinjected directly into hypothalamic nuclei such as the ARC of anesthetized rats using a micoinjector fitted to the stereotaxic apparatus or via a preimplanted guide cannula targeted to the ARC. The volume of injection should be <0.5 µL to avoid tissue damage.
3. The dose used has generally been 25 or 50 µg/injection (approx 4.0 to 8.0 nmol) for intraventricular administration and 250–500 ng for intrahypothalamic injections.
4. The duration and frequency of injections is important and is dependent on the stability of the antisense sequence (*see* **Note 3**).
5. When using the unmodified phosphodiester that is subject to rapid degradation in the brain, we administered three icv injections of 25 µg each within a 4-h period. This regimen produced no apparent toxic effects and effectively blocked the neuroendocrine actions of NPY *(21–23)*. The phosphorothioate form, on the other hand, produced toxicity even when injected only once a day *(24)*.
6. NPY antisense ODN has been continuously infused into the cerebral ventricle for 7 days via Alzet osmotic minipumps *(19)*. This mode of administration obviates the need for repeated injections; however, it did not enhance efficacy *(19)*.

### 3.3. NPY Antisense and Hormone Secretion

1. Analysis of NPY in the hypothalamus established that NPY gene expression and NPY levels increase on proestrus in association with the surge release of pituitary LH that induces ovulation in female cycling rats *(25,26)*. Similar changes in hypothalamic NPY activity were noted in association with the steroid-induced LH surge in ovariectomized (ovx) rats *(27)*.

2. To determine whether these increases in hypothalamic NPY activity represented *de novo* synthesis of NPY that was essential to elicit the LHRH and LH surges, we employed a 20-base unmodified phosphodiester NPY antisense sequence encompassing bases 74–93 of the NPY gene sequence *(21)*.

3. The antisense sequence (5'-CCA TTC GTT TGT TAC CTA GC-3') and a sense (5'-GCT AGG TAA CAA ACG AAT GG-3') and missense (5'-AAG GTA GAT ACT GGA ACC GA-3') sequence with the same G and C composition were dissolved in saline and injected into the lateral ventricle of ovx rats pretreated with estradiol and progesterone.

4. The sense, missense, or antisense ODNs (4 nmol/5 μL saline/injection) were injected three times at 2-h intervals via preimplanted guide cannulae. No untoward behavioral effects were noted; however, secretion of the LH surge was completely blocked in rats injected with the antisense ODN.

5. To confirm that the antisense ODN blocked mRNA translation, NPY levels were measured (*see* **Note 5**). NPY elevations in the median eminence-arcuate area (ME-ARC) and the medial and lateral preoptic areas that were characteristically seen in control rats were prevented in antisense ODN-injected rats. The specificity of the antisense ODN treatment was also affirmed by the ineffectiveness of the sense and missense sequences to prevent both the LH surge and augmentation of hypothalamic NPY levels.

6. The above NPY antisense sequence was also effective when injected into the third cerebroventricle (three injections of 4 nmol/3 μL saline each). It prevented both the increase in NPY levels in the ME-ARC and the LH surge induced by blockade of opiate receptors in estrogen-treated ovx rats *(22)*.

7. Three injections of NPY antisense ODN (4 nmol/3 μL saline each) at 2-h intervals into the third ventricle blocked the pulsatile release of LH in ovx rats *(23)*, demonstrating that the pulsatile pattern of endogenous NPY release *(28)* is critical for the generation of LHRH pulsatility. These results clearly demonstrated the efficacy of NPY antisense ODNs to block NPY synthesis and also established the usefulness of this molecular tool for analyzing the physiological role of this hypothalamic neuropeptide. No nonspecific effects of the sense strand or toxicity were noted in these acute experiments (*see* **Note 6**).

### 3.4. NPY Antisense Oligos and Appetite

1. We speculated that to block the behavioral effects of NPY in stimulating appetite, it may be necessary to inhibit ongoing NPY synthesis over a longer period (*see* **Note 7**).

2. To protect the ODN from rapid degradation in vivo, all bases of the NPY antisense sequence targeted to bases 74–93 were phosphorothioated.

3. Rats were injected once daily with this modified ODN (4 or 8 nmol/3 µL saline/ injection) into the third cerebroventricle via preimplanted guide cannulae *(24)*.

4. Body weights and 24-h consumption of chow were recorded. No effect on appetite was evident after the first injection; however, both doses of the antisense ODN reduced food intake during the second 24-h period by approx 50% ($p < 0.05$). The third injection of NPY antisense ODN almost completely inhibited food intake.

5. These results show that blockade of NPY synthesis in the ARC by icv administration of the antisense ODN causes a dramatic reduction in appetite, thus confirming the potent appetite-transducing effects of endogenous NPY.

6. Severe side effects were also noted in these experiments. Several of the antisense treated rats displayed huddling, tremors, and lack of grooming. The adverse effects were even more apparent in rats injected with the sense oligo, some of which also displayed seizures and succumbed. Thus, it is possible that the dramatic loss of appetite in this experiment may be a nonspecific result of the toxic effects of the phosphorothioated ODNs *(24)*.

7. Hulsey et al. *(29)* used a 5–10-fold lower dose of a phosphorothioated antisense sequence targeted to bases 78–95 (5'-CCC CAT TCG TTT GTT ACC-3'). Five micrograms of this antisense or a missense ODN (5'-TTA TTC CCC CAG TTT GCC-3') were injected daily into the lateral cerebroventricle for 7 days. The antisense ODN reduced cumulative food intake, meal size, and meal duration within 1 day, and this inhibition was maintained for 6 days. No evidence of sickness or taste aversion was reported with this lower dose. However, NPY levels in the PVN of the antisense and missense ODN-treated rats were similar, raising the possibility of nonspecific effects of the missense sequence. This reiterates the need for an additional saline-treated control group for studies with antisense ODNs (*see* **Note 2**).

8. Akabayashi et al. *(30)* employed a different form of antisense delivery. They injected a mixture of two unmodified antisense NPY oligos (total 0.5 µg/day) directly into the ARC for 4 days. One of the sequences encompassed bases 68–86 including the initiation codon of the signal peptide (5'-TTT GTT ACC TAG CAT CTG-3'), and the other corresponded to bases 207–224 encoding amino acids 18–23 of the mature NPY (5'-AGC GGA GTA GTA TCT GGC-3'). Only the antisense ODN treatment regimen caused a small reduction in ARC NPY levels and significantly reduced daily food intake by 20–30%. Although this phosphodiester ODN would be subject to rapid degradation in neurons, direct delivery into the ARC, where NPY-producing neurons are concentrated, may have contributed to its efficacy in reducing NPY levels and suppressing appetite for an extended period (*see* **Note 5**).

9. Schaffhauser et al. *(31)* reported that three intraventricular injections (50 µg each) over a 24-h period of this NPY antisense oligo (bases 68–86), with only the terminal bases phosphorothioated, prevented the fasting-induced increase in NPY

in the ARC and PVN and also reduced food intake (*see* **Note 3**). These effects were specific to the antisense-injected rats since in rats injected with the corresponding missense (scrambled) ODN, the fasting-induced increases in NPY levels and feeding were not altered compared with control saline-injected rats (*see* **Note 5**). No toxic effects were reported of the phophorothioate capped antisense and missense ODNs despite the relatively high dose (150 µg/24 h).

10. Dryden et al. *(19)* compared the relative efficacies of a NPY antisense ODN on the basis of structural modifications. A 20-base ODN encompassing bases 64–83 at the initiation codon (5'-GTT ACC TAG CAT CAT GGC GG-3') was infused continuously for 7 days via an Alzet osmotic minipump into the third cerebroventricle either as an unmodified phosphodiester, fully phosphorothioated or 3'-propyl modified. The unmodified sequence was undetectable in the hypothalamus and was ineffective in suppressing intake. The 3'-propyl-protected form, on the other hand, was widely dispersed in the hypothalamus but was only modestly effective in reducing food intake. The fully phosphorothioated form was also widely distributed and caused larger reductions in food intake and body weight. However, in this long-term infusion study, similar reductions were also seen in rats treated with the missense ODN; in agreement with the experience of other investigators, phosphorothioated ODN-injected rats (both antisense and missense) displayed signs of toxicity (listlessness, hunched posture, no grooming). The nonselective effects of the 3'-propyl and phosphorothioated forms along with the toxicity caused by the latter modification led the investigators to question the overall validity of this approach (*see* **Notes 3** and **7**). However, the mode of delivery and length of treatment may have contributed to the nonspecific effects in this study.

## 1. Notes

1. Any of the published sequences for NPY antisense ODNs may be used.
2. Control groups must include a saline (vehicle) control and a missense sequence group. Substitution of just a few nucleotides on the antisense sequence is adequate to prevent hybridization of the missense sequence.
3. The recommended base structure modification that reduces in vivo degradation and increases half-life of the ODN without producing toxicity is end capping, i.e., phosphorothioate modification of one or two terminal bases, especially at the 3'-end.
4. NPY ODNs must be delivered centrally. Both the lateral or third cerebral ventricle route of delivery are effective.
5. The specificity of the antisense treatment must be confirmed by measuring NPY levels in the hypothalamus.
6. The neuroendocrine effects of hypothalamic NPY are rapidly inhibited (within hours) by centrally administered antisense ODNs. Unmodified phosphodiesters can be used for these short-term studies.
7. Blocking the behavioral effects of NPY by antisense ODNs requires a longer period of action. Terminal phosphorothioated sequences are suitable for studies that require several days of ODN action.

## References

1. Chronwall, B. M. (1985) Anatomy and physiology of the neuroendocrine arcuate nucleus. *Peptides* **2,** 1–11.
2. Kalra, S. P. and Kalra, P. S. (1996) Nutritional infertility: the role of the interconnected hypothalamic neuropeptide Y-galanin-opioid network. *Front. Neuroendocrinol.* **17,** 371–401.
3. Clark, J. T., Kalra, P. S., Crowley, W. R., and Kalra, S. P. (1984) Neuropeptide Y and human pancreatic polypeptide stimulate feeding behavior in rats. *Endocrinology* **115,** 427–429.
4. Stanley, B. G., Chin, A. S., and Leibowitz, S. F. (1985) Feeding and drinking elicited by central injection of neuropeptide Y: evidence for a hypothalamic site(s) of action. *Brain Res. Bull.* **14,** 521–524.
5. Ciofi, P., Fallon, J. H., Croix, D., Polak, J. M., and Tramu, G. (1991) Expression of neuropeptide Y precursor-immunoreactivity in the hypothalamic dopaminergic tubero-infundibular system during lactation in rodents. *Endocrinology* **128,** 823–834.
6. Smith, M. S. (1993) Lactation alters neuropeptide-Y and proopiomelanocortin gene expression in the arcuate nucleus of the rat. *Endocrinology* **133,** 1258–1265.
7. Williams, G., Gill, J. S., Lee, Y. C., Cardoso, H. M., Okpere, B. E., and Bloom, S. R. (1989) Increased neuropeptide Y concentrations in specific hypothalamic regions of streptozocin-induced diabetic rats. *Diabetes* **38,** 321–327.
8. Sahu, A., Sninsky, C. A., Kalra, P. S., and Kalra, S. P. (1990) Neuropeptide-Y concentration in microdissected hypothalamic regions and in vitro release from the medial basal hypothalamus-preoptic area of streptozotocin-diabetic rats with and without insulin substitution therapy. *Endocrinology* **126,** 192–198.
9. Sahu, A., Kalra, P. S., and Kalra, S. P. (1988) Food deprivation and ingestion induce reciprocal changes in neuropeptide Y concentrations in the paraventricular nucleus. *Peptides* **9,** 83–86.
10. Brady, L. S., Smith, M. A., Gold, P. W., and Herkenham, M. (1990) Altered expression of hypothalamic neuropeptide mRNAs in food-restricted and food-deprived rats. *Neuroendocrinology* **52,** 441–447.
11. Sahu, A., White, J. D., Kalra, P. S., and Kalra, S. P. (1992) Hypothalamic neuropeptide Y gene expression in rats on scheduled feeding regimen. *Brain Res. Mol. Brain Res.* **15,** 15–18.
12. Crowley, W. R. and Kalra, S. P. (1987) Neuropeptide Y stimulates the release of luteinizing hormone-releasing hormone from medial basal hypothalamus in vitro: modulation by ovarian hormones. *Neuroendocrinology* **46,** 97–103.
13. Whitesell, L., Geselowitz, D., Chavany, C., Fahmy, B., Walbridge, S., Alger, J. R., et al. (1993) Stability, clearance, and disposition of intraventricularly administered oligodeoxynucleotides: implications for therapeutic application within the central nervous system. *Proc. Natl. Acad. Sci. USA* **90,** 4665–4669.
14. Allen, J., Novotny, J., Martin, J., and Heinrich, G. (1987) Molecular structure of mammalian neuropeptide Y: analysis by molecular cloning and computer-aided comparison with crystal structure of avian homologue. *Proc. Natl. Acad. Sci. USA* **84,** 2532–2536.

15. McCarthy, M. M., Brooks, P. J., Pfaus, J., Brown, H. E., Flanagan, L. M., Schwartz-Giblin, S.,et al. (1993) Antisense technology in behavioral neuroscience. *Neuroprotocols* **2,** 67–76.

16. Gao, W. Y., Han, F. S., Storm, C., Egan, W., and Cheng, Y. C. (1992) Phosphorothioate oligonucleotides are inhibitors of human DNA polymerases and RNase H: implications for antisense technology. *Mol. Pharmacol.* **41,** 223–229.

17. Wahlestedt, C. (1994) Antisense oligonucleotide strategies in neuropharmacology. *Trends Pharmacol. Sci.* **15,** 42–46.

18. Russell, D., Widnell, K. L., and Nestler, E. J. (1996) Antisense oligonucleotides: new tools for the study of brain function. *Neuroscientist* **2,** 79–82.

19. Dryden, S., Pickavance, L., Tidd, D., and Williams, G. (1998) The lack of specificity of neuropeptide Y (NPY) antisense oligodeoxynucleotides administered intracerebroventricularly in inhibiting food intake and NPY gene expression in the rat hypothalamus. *J. Endocrinol.* **157,** 169–175.

20. Neckers, L. M., Geselowitz, D., Clavany, C., Whitesell, L., and Bergen, R. (1995) Antisense efficacy: Site-restricted in vivo and *ex vivo* models, in *Methods in Molecular Medicine: Antisense Therapeutics*, vol. I (Agrawal, S., ed.), Humana, Totowa, NJ, pp. 47–56.

21. Kalra, P. S., Bonavera, J. J., and Kalra, S. P. (1995) Central administration of antisense oligodeoxynucleotides to neuropeptide Y (NPY) mRNA reveals the critical role of newly synthesized NPY in regulation of LHRH release. *Regul. Pept.* **59,** 215–220.

22. Xu, B., Sahu, A., Kalra, P. S., Crowley, W. R., and Kalra, S. P. (1996) Disinhibition from opioid influence augments hypothalamic neuropeptide Y (NPY) gene expression and pituitary luteinizing hormone release: effects of NPY messenger ribonucleic acid antisense oligodeoxynucleotides. *Endocrinology* **137,** 78–84.

23. Xu, B., Pu, S., Sahu, A., Kalra, P. S., Hyde, J. F., Crowley, W. R., et al. (1996) An interactive physiological role of neuropeptide Y and galanin in pulsatile pituitary luteinizing hormone secretion. *Endocrinology* **137,** 5297–5302.

24. Kalra, P. S., Dube, M. G., and Kalra, S. P. (1999) Effects of centrally administered antisense oligodeoxynucleotides on feeding behavior and hormone secretion, in *Methods in Enzymology*, vol. 314B, (Phillips, M. I, ed.), Academic, San Diego, CA, pp. 184–200.

25. Sahu, A., Jacobson, W., Crowley, W. R., and Kalra, S. P. (1989) Dynamic changes in neuropeptide Y concentrations in the median eminence in association with preovulatory luteinizing hormone (LH) release in the rat. *J. Neuroendocrinol.* **1,** 83–87.

26. Sahu, A., Crowley, W. R., and Kalra, S. P. (1995) Evidence that hypothalamic neuropeptide Y gene expression increases before the onset of the preovulatory LH surge. *J. Neuroendocrinol.* **7,** 291–296.

27. Sahu, A., Crowley, W. R., and Kalra, S. P. (1994) Hypothalamic neuropeptide-Y gene expression increases before the onset of the ovarian steroid-induced luteinizing hormone surge. *Endocrinology* **134,** 1018–1022.

28. Sahu, A., Phelps, C. P., White, J. D., Crowley, W. R., Kalra, S. P., and Kalra, P. S. (1992) Steroidal regulation of hypothalamic neuropeptide Y release and gene expression. *Endocrinology* **130,** 3331–3336.
29. Hulsey, M. G., Pless, C. M., White, B. D., and Martin, R. J. (1995) ICV administration of anti-NPY antisense oligonucleotide: effects on feeding behavior, body weight, peptide content and peptide release. *Regul. Pept.* **59,** 207–214.
30. Akabayashi, A., Wahlestedt, C., Alexander, J. T., and Leibowitz, S. F. (1994) Specific inhibition of endogenous neuropeptide Y synthesis in arcuate nucleus by antisense oligonucleotides suppresses feeding behavior and insulin secretion. *Brain Res. Mol. Brain Res.* **21,** 55–61.
31. Schaffhauser A. O., Sricker-Krongrad, A., Brunner, L., Cumin, F., Gerald, G., Whitebread, S., et al. (1997) Inhibition of food intake by neuropeptide Y5 antisense oligodeoxynucleotides. Diabetes 46, 1792–1798.

# IV

## TECHNIQUES TO STUDY TISSUE DISTRIBUTION OF mRNA

# 13

# Localization of Y-Receptor Subtype mRNAs in Rat Brain by Digoxigenin Labeled *In Situ* Hybridization

## Rachel Parker and Herbert Herzog

## 1. Introduction

*In situ* hybridization histochemistry (ISHH), first described in 1969 *(1,2)*, allows a specific complementary RNA species to be detected directly at its site of expression, revealing its cellular localization and relative abundance. The utilization as a label of digoxigenin *(3)*, which is not endogenous to mammalian tissue, provides a sensible alternative to radiolabels, having obvious inherent advantages (e.g., safety, speed, and higher cellular resolution), yet provides comparable sensitivity *(4–7)*. The method basically includes the following six steps: (1) probe labeling (the cloning techniques needed to produce suitable vector templates for cDNA and riboprobe synthesis are not covered in this chapter); (2) tissue preparation; (3) prehybridization tissue treatment; (4) hybridization; (5) posthybridization washing; and (6) signal detection.

The Y-receptor gene family represents the largest and most complex family of G-protein-coupled receptors for any known peptide *(8)*. When activated by their ligands, neuropeptide Y (NPY), peptide YY (PYY), and pancreatic polypeptide (PP) *(9)*, they can modulate a wide variety of important physiological events, including food and water intake, circadian rhythms, anxiety, memory, nociception, hormone release, and blood pressure regulation *(10–12)*. Four functional Y-receptor subtypes, Y1, Y2, Y4, and Y5, have been cloned and characterized *(13–16)*. These Y-receptor subtypes show low primary amino acid sequence identity (as low as 30%), yet surprisingly exhibit very similar pharmacology *(16–18)*. Chimeric peptides and fragments of the NPY peptide family have been used to distinguish between different subtypes. Unfortunately, the selective pharmacology of several of these analogs has now been

From: *Methods in Molecular Biology, vol. 153: Neuropeptide Y Protocols*
Edited by: A. Balasubramaniam © Humana Press Inc., Totowa, NJ

invalidated by the cloning of other Y-receptor family members *(8)*. This lack of subtype-specific ligands, together with the possible existence of other subtypes, has made it very difficult to determine the specific roles of each subtype in physiological processes.

To investigate the physiological significance of the different Y-receptor subtypes in various central nervous system (CNS) functions, while avoiding possible ambiguity created by a limited pharmacological approach, we have comprehensively mapped and compared, at a cellular level, the differential distribution of Y1-, Y2-, Y4-, and Y5-receptor subtype mRNA expression. This was carried out within consecutive sections throughout the rat CNS by ISHH, using digoxigenin-labeled riboprobes. The protocols described here have been successfully applied to an investigation of the regional distribution of Y-receptor subtype mRNAs in rat brain *(19,20)*. This should provide a useful framework to other applications.

## 2. Materials
### 2.1. Chemicals and Solutions

It is important to take precautions against RNase contamination when making the following solutions (*see* **Note 1**).

1. Agarose (1%)/formaldehyde solution: for 100 ml, dissolve 1 g agarose (Ultrapure, Life Technologies, Renfrewshire, Scotland; BRL-540-5510UA) in 72.1 mL RNase-free distilled $H_2O$, cool to approx 60°C, and then add 10 mL of 10X 3-(*N*-morpholino)-propane sulfonic acid (MOPS) running buffer. In a fume hood, add 17.7 mL of formaldehyde (37% solution), to give a final concentration of 2.2 *M*. Allow to cool before pouring the gel (agarose sets at approx 45°C).
2. Alkaline phosphate buffer: 100 m*M* NaCl, 5 m*M* $MgCl_2$, 100 m*M* Tris-HCl, pH 9.5.
3. Antibody solution: for *in situ* hybridizations add sufficient anti-digoxigenin-alkaline phosphatase-conjugated antibody (Boehringer Mannheim, Germany; #1093 274) at 0.75 U/µL to blocking solution to give a final antibody concentration of at least 1:500 (we use 2 µL antibody at 0.75 U/µL/mL of the blocking solution); for blots use a dilution of 1:5000 (we use 2 µL antibody at 0.75 U/µL/ 10 mL of the blocking solution). Make fresh.
4. Aquamount mounting medium.
5. BCIP-NBT color substrate solution: dissolve 0.1 g of 5-bromo-4-chloro-3-indolyl phosphate disodium salt (BCIP: #B8503, Sigma, Castle Hill, NSW, Australia) in 2 mL of 100% dimethylformamide. Dissolve 0.2 g of nitro blue tetrazolium chloride (NBT: Sigma #N6876) in 4 mL of 70% dimethylformamide. Mix 1.88 mL of BCIP solution, 3.75 mL of NBT solution, 2.5 mL of dimethylformamide, and 1.87 mL of distilled $H_2O$ to obtain 10 mL stock solution containing 9.4 mg/mL BCIP and 18.75 mg/mL NBT in 70% dimethylformamide. Store in a dark glass bottle at 4°C.

6. Blocking buffer: blocking solution containing 0.3% Triton X-100. Make fresh.

7. Blocking solution: 2% skim milk powder in 1X Tris-buffered saline (TBS). Heat in microwave (40% power) to dissolve, and filter through 3M Whatman prefolded filter paper. Store at 4°C for up to 1 week.

8. Color substrate solution: add 20 µL BCIP-NBT color substrate solution/mL alkaline phosphate buffer (to give 188 µg/mL BCIP; 375 µg/mL NBT). Add 1 $M$ levamisole solution to give a final concentration of 1 m$M$. Vortex to mix, filter through a 0.2-µm filter into a foil wrapped tube, and keep in the dark until use. Make fresh.

9. Denhardt's solution (50X): 1% (w/v) Ficoll type 400 (Sigma #F2637), 1% (w/v) polyvinylpyrrolidene (PVP; Sigma #P5288), 1% (w/v) bovine serum albumin (fraction V; Sigma #A7030), made up in RNase-free distilled $H_2O$. Filter through a Millipore 0.2-µm filter, and store in 10-mL aliquots at –20°C.

10. DePX mounting medium.

11. Diethyl pyrocarbonate (DEPC)-treated distilled $H_2O$ (0.05%): in a fume hood, add 0.5 mL DEPC (Sigma #D5758; stock stored at 4°C; very toxic) per L of distilled $H_2O$, shake well, then stand for 20 min, shake again, and set at room temperature for >2 h, but <24 h, and then autoclave for 35–40 min to destroy excess DEPC.

12. Digoxigenin-labeled control RNA (100 ng/µL; Boehringer Mannheim #158746).

13. Dithiothreitol (DTT): make a 1 $M$ solution of DTT (Sigma #D9779) in RNase-free distilled $H_2O$, and store at –20°C in an autoclaved microcentrifuge tube. This solution oxidizes quickly at room temperature.

14. EDTA: Make a 0.5 $M$ solution of EDTA (Sigma #E5134) in RNase-free distilled $H_2O$, stir, and dissolve by adjusting to pH 8.0 with NaOH (10 $M$), and then autoclave.

15. Eosin Y in 80% ethanol: dissolve 1% eosin Y (Sigma E4382; Sigma, Castle Hill, NSW, Australia) in 80% ethanol solution. Filter twice through Whatman no. 1 filter paper prior to use.

16. Ethanol: absolute, analytical grade.

17. Ethanolamine solution: 0.15 $M$ NaCl (Sigma #S3014), 0.1 $M$ triethanolamine (Merck, Victoria, Australia; #BDH 10370), and 0.25% (v/v) acetic anhydride (Merck #BDH 100026Q). Make fresh.

18. Ethidium bromide (EtBr) solution: make a 10 mg/mL solution of EtBr (Sigma #E7637) in distilled $H_2O$; store at 4°C in a sealed, dark bottle protected against light (very hazardous/carcinogen).

19. Formamide: mix 100 mL formamide/10 g mixed-bed ion exchange resin 20–50 mesh (Bio-Rad, NSW, Australia; #AG 501-X8) for 30–60 min at room temperature, in a fume cupboard, filter twice through Whatman no. 1 filter paper, dispense into aliquots, and store at –20°C.

20. Formaldehyde: 37% solution (12.3 $M$); Sigma #F8775; hazardous; store in chemical box.

21. Histoclear: use in a fume hood.

22. Hybridization solution (2X) for ISHH: dissolve 20% (w/v) dextran sulphate (Sigma #D8906; helps probe anneal) in 10 mL RNase-free distilled $H_2O$,

heated to 60°C, and then add 4 mL of 20X SSPE (final conc. 2X), 800 µL of 50X Denhardt's (final conc. 1X) 80 µL of 50 mg/mL yeast tRNA (final conc 100 µg/mL; stabilizes the probe and is a nonspecific blocker). Make up volume to 20 mL with RNase-free distilled $H_2O$. Mix well, taking care not to denature the bovine serum albumin (BSA; NB solutions containing BSA cannot be autoclaved), and then store in 500-µL aliquots at –20°C for up to 6 months.

23. Levamisole (1 *M*): add 240 mg levamisole (Sigma #L9756) per mL distilled $H_2O$, dissolve, and store at 4°C.

24. Mayer's hematoxylin solution: (Sigma #MHS-16; Sigma) filter twice through Whatman no. 1 filter paper before use.

25. MOPS running buffer (X10): for 1 L, dissolve 4.1 g sodium acetate (Sigma #S2889) in 700 mL RNase-free distilled $H_2O$, add 41.9 g MOPS (Sigma #M8899) adjust pH to 7.0 with 10 *M* NaOH, then add 10 mL of 0.5 *M* EDTA, pH 8.0, make up volume with RNase-free distilled $H_2O$, autoclave, and store at 4°C.

26. Nembutal anesthetic: sodium pentabarbitone (60 mg/mL; Rhone Merieux).

27. Nucleotide labeling mix (Boehringer Mannheim).

28. Nylon membrane (Amersham HybondN+, #RPN 203B).

29. Paraformaldehyde (4% w/v) in 0.1 *M* phosphate-buffered saline (PBS): For 1 L, warm 500 mL distilled $H_2O$ to 65°C in a fume cupboard, and slowly add 40 g paraformaldehyde (Merck #294474L). Add approx 200 µL NaOH (10 *M*), and stir vigorously to facilitate dissolving of paraformaldehyde (takes <1 h). Cool, add 500 mL 0.2 *M* PBS, and adjust pH to 7.5. This solution can be kept for 1 month at 4°C, after which it may polymerize. It will also polymerize if autoclaved (very hazardous chemical).

30. PBS, 0.1 *M*): 10 m*M* $Na_2HPO_4$ (Sigma #S3264), 3 m*M* KCl (Sigma #P9541), 1.8 m*M* $KH_2PO_4$ (Sigma #P9791), 140 m*M* NaCl (Sigma #S3014), adjust pH to 7.4, and autoclave. This solution can be kept for 2–3 weeks at 4°C.

31. PBS/glycine solution: 0.1 *M* glycine (Sigma #G7032) in 0.1 *M* PBS. Make fresh.

32. Poly-D-lysine/fibronectin solution: make a 1 in 20 dilution of poly-D-lysine in 0.1 *M* PBS. Make a 1 in 25 dilution of fibronectrin in 0.1 *M* PBS. Mix 1 part poly-D-lysine and 1 part fibronectrin.

33. Protinase K solution: 50 m*M* Tris-HCl, pH 7.5, 5 m*M* EDTA, containing 10 mg/mL proteinase K (Boehringer Mannheim #745723). Store in 2-µL aliquots at –20°C and use at 5 µg/mL in 50 m*M* Tris-HCl, pH 7.5 containing 5 m*M* EDTA solution.

34. RNA loading dye: 0.4% bromophenol blue (Sigma #B5525), 0.4% xylene cyanol (Sigma #X4126), 1 m*M* EDTA (pH 8), 50% glycerol (Sigma #G5516). Store stock at 4°C.

35. RNA markers: 0.24–9.5-kb ladder (Gibco-BRL #15626). Store at –20°C.

36. RNA polymerases: SP6, T7, or T3 polymerases supplied with 10X transcription buffer and 100 m*M* DTT (Promega #sP1085, P2075, and P2083). Store at –20°C or at –70°C for long term storage.

37. RNase-free DNase: (Promega #M6101). Store at –20°C.

38. RNaseA solution buffer: 10 m$M$ Tris-HCl, pH 7.6, 0.5 $M$ NaCl (Sigma #S3014), 1 m$M$ EDTA (Sigma #E5134; Sigma). Adjust pH to 8.0, autoclave, and store at 4°C.
39. RNaseA solution: dissolve ribonuclease A (RNase A; bovine pancreas; Sigma #R5503; stock stored –20°C) at 25 mg/mL in autoclaved distilled H$_2$O (which has not been DEPC treated). Store aliquots at –20°C. When ready to use, boil an aliquot for 1 min to remove DNase activity, and then dilute this aliquot in RNaseA solution buffer to give a final concentration of 25 µg RNase/mL buffer.
40. RNasin: Promega #N2511) 20–40 U/µL. Store at –20°C.
41. ssDNA (from salmon sperm, sonicated): Sigma #D9156; stored as 10 mg/mL solution at –20°C. Just prior to use, denature this solution at 100°C for 5 min, and rapidly cool on ice.
42. Sephadex slurry: equilibrate Sephadex G25 powder (Sigma #G-25-80) in 1X TE buffer, pH 8.0 overnight. Store slurry at 4°C.
43. SSPE (20X): 3 $M$ NaCl (Sigma #S3014), 0.2 $M$ NaH$_2$PO$_4$ (Sigma #S3139), 20 m$M$ EDTA. Adjust to pH 7.4 with 10 $M$ NaOH, and autoclave.
44. Standard sodium citrate solution (SSC; 20X): 3 $M$ sodium chloride (Sigma #S3014), 0.3 $M$ Tri-sodium citrate (Sigma C8532; Sigma, Castle Hill, NSW, Australia) in RNase-free dH$_2$O. Adjust to pH 7.0, autoclave (will keep 2–3 weeks at room temperature).
45. TBS (10X): 1.5 $M$ NaCl, 1 $M$ Tris. Adjust pH to 7.5 with HCl.
46. TE buffer (1X): 10 m$M$ Tris, 1 m$M$ EDTA. Adjust to pH 8.0 with 10 $M$ NaOH and autoclave.
47. Tris (1 $M$): add 6.35 g Trizma hydrochloride (Sigma #T7149) to 1.18 g Trizma base (Sigma #T8524) in 50 mL of autoclaved distilled H$_2$O, and then autoclave to give a solution of pH 7.5 at 25°C. We do not use DEPC-treated distilled H$_2$O because if there is any residual DEPC present, it will react with Tris.
48. tRNA (50 mg/mL): dissolve 100 mg type X-SA baker's yeast tRNA (Sigma #R8759; 2000 U/100 mg) in 2 mL RNase-free distilled H$_2$O, vortex to mix, and store at –20°C.
49. Xylene: use in a fumehood (carcinogenic).

## 2.2. Equipment

1. Gel apparatus and power supply.
2. Chamber slides (Nalge Nunc, Naperville, IL).
3. Disposable microtome blades: stainless steel (Feather Safety Razor, Japan).
4. Hybaid OmniSlide System: OmniSlide thermal cycler and wash modules with slide racks (Hybaid, Middlesex, UK).
5. Microscope.
6. Microscope coverslips (Merck #BDH 406/0188/42) and slides (Merck #BDH 406/0169/02 or BDH 406/0179/00).
7. Microtome and water bath (Leica, NSW, Australia).
8. UV transilluminator.
9. Wax pen: Zymed, San Francisco, CA.
10. Whatman (Clifton, NJ) chromatography paper (3MM).

## 3. Methods

It is important to take precautions against RNase contamination when carrying out the following protocols (*see* **Note 1**).

### 3.1. Probe Labeling

### 3.1.1. Spun Column Preparation

1. Plug a sterile 1-mL syringe barrel with glass microfiber filter paper, packing it in place with the syringe plunger.
2. Slowly pipet 1 mL of Sephadex slurry into the column, taking care to avoid air bubbles.
3. Spin column at >1100$g$ for 4 min.
4. Repeat **steps 2** and **3** until the column contents stabilize at 1 mL of Sephadex G25.
5. Add a known volume (100 μL) of 1X TE buffer to the top of the Sephadex bed, place the column in a microcentrifuge tube, and repeat **step 3**. The volume recovered should be equal to that added to the top of the column.
6. Repeat **step 5** twice.
7. Add 100 μL of 1X TE buffer to the top of the Sephadex bed, seal the column with Parafilm to prevent evaporation, and store upright at 4°C until ready to use (*see* **Note 2**).

### 3.1.2. In Vitro Transcription Labeling
### of Riboprobes with Digoxigenin

1. In an autoclaved microcentrifuge tube on ice, add in order:

|  |  | (final) |
|---|---|---|
| 10X transcription buffer (supplied with enzyme) | 2 μL | 1X |
| 10X nucleotide labeling mix | 2 μL | 1X |
| Linearized cDNA template in distilled H$_2$O (*see* **Note 3**) | 12 μL | 1 μg/reaction |
| 100 m$M$ DTT (supplied with enzyme; see **Note 4**) | 1 μL | 10 m$M$ |
| rRNasin (20-40 U/μL) | 1 μL | 2–4 U/μL |
| RNA polymerase (20 U/μL) (*see* **Note 5**) | 2 μL | 2 U/μL |
| Total volume | 20 μL | |

2. Flick the tube to gently mix, spin for 2–3 s to remove air bubbles, which may inhibit enzyme activity, and incubate at 37°C for 2 h.
4. Add 1 U of RNase-free DNase/μg DNA, and incubate at 37°C for 15 min to destroy the DNA template.
5. Stop the reaction with 0.8 μL of 0.5 $M$ EDTA, pH 8.0, and make the volume up to 100 μL with RNase-free distilled H$_2$O.
6. Purify the riboprobe through a spin column (as prepared in **Subheading 3.1.1.**): prespin the spun column for 4 min at 1100$g$ and discard the run-through buffer, place the spin column in a fresh collecting microcentrifuge tube, apply the sample (100 μL) to the center of the top of the Sephadex bed in the spin column, and spin for 4 min at 1100$g$. The unincorporated label will remain in the column, while the labeled probe runs through.

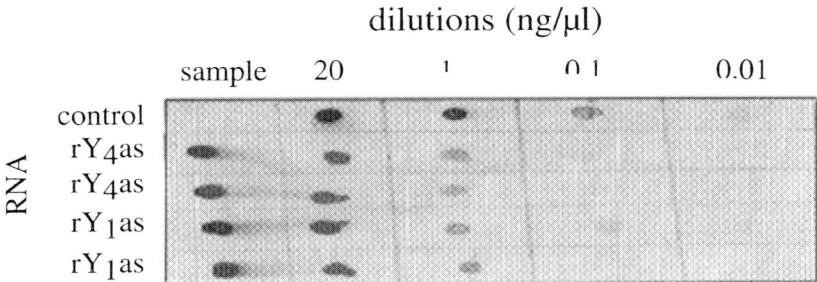

Fig. 1. Assessment of digoxigenin-labeled riboprobe concentration by dot-blot procedure. The concentration of digoxigenin-labeled riboprobe is estimated to be approx 10 ng/μL (rows 2 and 3; rat $Y_4$ antisense riboprobes, rows 4 and 5; rat $Y_1$ antisense riboprobes), when serial dilutions are compared against serial dilutions of the control digoxigenin-labeled RNA (top row).

7. Retain the labeled probe in the collecting tube, the final volume of which should be of equal volume to that applied on the column. Store at –20°C in 10-μL aliquots (for up to a year), and thaw slowly on ice to use.

### 3.1.3. Assessment of Digoxigenin-Labeled Riboprobe Concentration

#### 3.1.3.1. DOT-BLOT PROCEDURE

1. On ice, carry out serial dilutions of digoxigenin-labeled control RNA (100 ng/μL) in distilled $H_2O$ to obtain final concentrations of 20 ng/μL, 1 ng/μL, 100 pg/μL, and 10 pg/μL. Make serial dilutions of riboprobe to obtain final dilutions of 1 in 5, 1 in 100, 1 in 1000 and 1 in 10,000.
2. Spot 1 μL of each dilution onto nylon membrane, one row for control dilutions and separate rows for each riboprobe.
3. Air-dry the membrane for 30 min by placing between two folds of filter paper.

#### 3.1.3.2. PROBE DETECTION

1. Soak the membrane with agitation in antibody solution (containing a 1:5000 dilution of antidigoxigenin antibody) for 30 min at room temperature.
2. Wash the membrane in 1X TBS with agitation, and twice for 15 min each.
3. Rinse the membrane in alkaline phosphatase buffer for approximately 2 min.
4. Incubate the membrane for approximately 2 h in the dark with freshly prepared solution of 200 μL BCIP-NBT color substrate solution/10 mL alkaline phosphatase buffer.
5. When the color has developed sufficiently, stop the reaction by washing the membrane in water for 5 min, and dry it on two folds of filter paper in the dark, before assessing the riboprobe concentration against the control dilutions. We routinely obtain a concentration of 10 ng/μL labeled riboprobe/labeling reaction (*see* **Fig. 1**).

## 3.1.4. Assessment of Riboprobe Quality and Molecular Weight

### 3.1.4.1. RNA GEL FRACTIONATION

1. Soak gel apparatus in DEPC-treated distilled $H_2O$ for at least 1 h (If possible, it is better to dedicate a gel tray and wells purely for RNA work to minimize the risk of RNase contamination and therefore degradation of RNA.)
2. In a fume hood, pour a 1% agarose-formaldehyde gel solution (no more than 0.50 cm thick) into an RNase-free gel tray. Allow to settle (takes 15–30 min), and then equilibrate with 1X MOPS running buffer for at least 15 min.
3. Prepare fresh loading solution by mixing 100 µL of 10X MOPS running buffer with 175 µL of formaldehyde and 500 µL deionized formamide.
4. For each RNA sample, dilute 2 µL of RNA (or at least 2 ng) with loading solution to give a final volume of 10 µL, and gently flick to mix. Similarly prepare 10 µL of RNA marker solution, containing 3 µg RNA markers diluted in loading solution.
5. Incubate the diluted samples at 65°C for 15 min to denature the RNA, and then rapidly cool them on ice to keep the RNA denatured.
6. Add 1 µL of RNA loading dye to each 10 µL of diluted sample and carefully pipet the aliquots into separate wells of the set gel, keeping track of the order in which the samples and markers are loaded. Take care not to pierce the bottom of the gel or to expel the last drop of solution from the pipet tip, as this may cause air bubbles, which could displace the solution out of the well. Expedite this step to minimize sample diffusion.
7. Run the gel at 60 V for up to 3 h or until the dye front is at least two-thirds along the gel.
8. Stain the markers with approximately 4 µL EtBr solution (10 mg/mL)/100 mL running buffer for 20 min. Destain overnight in RNase-free distilled $H_2O$ at 4°C to remove formaldehyde and excess EtBr. Visualize the RNA on a UV transilluminator, and photograph the bands against a millimeter ruler. Plot a linear standard molecular weight calibration curve of the log of the molecular weight against band migration distance (mm) from the well.
9. Trim off any excess gel from around the edges, and notch a corner to identify the gel orientation.
10. Soak the gel for 20 min in 0.05 $N$ NaOH, rinse in RNase-free distilled $H_2O$, soak for 45 min in 20X SSC and immediately transfer the RNA to a membrane by Northern blotting.

### 3.1.4.2. NORTHERN BLOTTING

1. Transfer RNA to nylon membrane by capillary transfer in 10X SSC for approximately 18 h.
2. Place the blotted membrane on filter paper soaked in 0.05 $M$ NaOH for 5 min to fix the RNA.
3. Briefly rinse the membrane in 2X SSC to remove any agarose, and air-dry for 30 min between two folds of filter paper.
4. Detect the riboprobe bands as in **Subheading 3.1.3.2.**

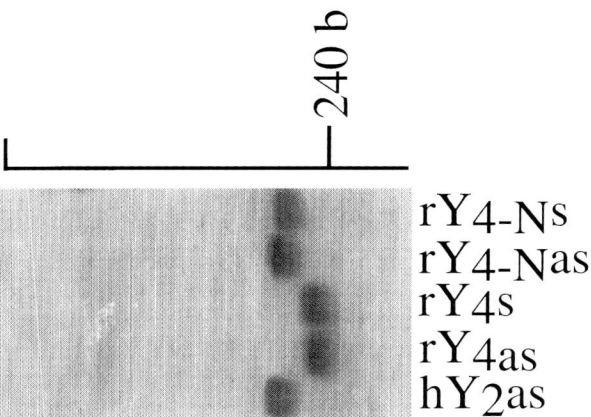

Fig. 2. Assessment of riboprobe quality and molecular weight by Northern blot. Each cDNA construct produces a riboprobe band of expected size, indicating successful in vitro transcription. hY2 corresponds to a Y2-receptor antisense riboprobe generated from cDNA corresponding to base sequence 838–1177 of the human Y2-receptor sequence (where 1 represents the first base of the translation initiation codon), which shows 100% identity to the rat sequence in this region *(14)*, subcloned into a pcDNA3 vector (Invitrogen, Carlsbad, CA). Two sets of Y4 antisense and sense riboprobes were generated: rY4as (antisense) and rY4s (sense) corresponding to base sequence 820–1146, and rY4-Nas (antisense) and rY4-Ns (sense) corresponding to base sequence 1–565 (**ref. *15***; where 1 represents the first base of the translation initiation codon), subcloned into pBluescript plasmid vectors (Stratagene). Each cDNA sequence we have used to generate riboprobes for the Y1-, Y2-, Y4-, and Y5 receptor subtypes has 100% identity with the corresponding receptor gene sequence, but exhibits <60% identity with any other known sequence.

5. Measure the migration distance (in mm) of detected bands on the membrane from the top of the marked well position. Calculate the molecular weight of these bands, using the calibration graph obtained in **Subheading 3.1.3.1. (step 10)**.
6. A good-quality probe will produce bands of the expected length, indicating successful in vitro transcription (*see* **Fig. 2**). Poor-quality probes will yield a smear of smaller sized products due to degradation or poor quality of template cDNA, or give bands of higher molecular weight than expected, indicating incomplete linearization of the template.

## 3.2. Tissue Collection and Section Cutting

To minimize RNA degradation and maximize signal detection, it is critical that RNase-free conditions be maintained and tissue be collected rapidly, and stored and fixed correctly.

1. Transcardiacally perfuse the rat with 4% paraformaldehyde in 0.1 *M* PBS under Nembutal anesthesia (sodium pentabarbitone, 60 mg/kg, ip) to fix the brain rapidly before removal.
2. Postfix the brain for at least 2 h with 4% paraformaldehyde solution in 0.1 *M* PBS.
3. Process the tissue for wax embedding. These wax blocks can then be stored (4°C) indefinitely.
4. On a microtome, collect individual 5-μm-thick sections by floating from a water bath at 37°C onto RNase-free poly-L-lysine-subbed or positively charged slides (Merck BDH superfrost plus microscope slides). We usually mount 2 brain sections/slide.
5. Bake the sections at 60°C for 2 h to melt the wax, allow them to cool, and store in sealed slide boxes at 4°C. Sections should be used in ISHH as soon as possible.
6. Sections can be taken for histological staining periodically to check section quality and assess the anatomical detail and stereotaxic location of areas of interest for *in situ* hybridization. For rat brain tissue we routinely use hematoxylin and eosin as follows (it may be necessary to adjust the staining time depending on the age of the stains and the condition of the tissue sections. For example, eosin will be absorbed very quickly when the sections are of good quality):
   a. Mayer's hematoxylin  20 s (a basic blue dye for nucleic acids)
   b. Distilled $H_2O$        5 s
   c. Alkaline $H_2O$         30 s  (1 drop of conc. ammonia in approx 300 mL distilled $H_2O$)
   d. Distilled $H_2O$        30 s
   e. 70% EtOH             1 min (needed because alcohol-based eosin is used)
   f. 1% eosin Y           Approx 5 s (stains cytoplasm a pale pink/orange)
   g. Acid alcohol (70%)   $\geq$15 s (this takes out excess eosin)
   h. 95% EtOH             2 × 1 min
   i. 100% EtOH            2 × 1 min
   j. Histoclear           2 × 2 min
   k. Immediately mount the sections in DePX mounting medium and gently coverslip with the aid of forceps to avoid air bubbles. (Glass coverslips can be removed from the sections at a later date by soaking the slides overnight in xylene to dissolve the DePX mounting medium.)

## 3.3. Prehybridization

The sensitivity of the method may be increased in several ways by employing one or many steps, depending on the nature of the tissue source, to (1) maintain RNA integrity (tissue fixation and use of RNase inactivators); (2) help reduce nonspecific background (delipidation, acetylation and prehybridization with hybridization solution minus probe); and (3) improve probe access (wax removal and proteinase K treatment).

1. Sterilize all prehybridization containers and hybridization boxes (*see* **Note 1**).

2. Stack the slides to be used in a slide rack. Return the remaining slides to the 4°C as soon as possible. We use a Hybaid OmniSlide system and find that up to 20 slides is a manageable number to assay at once.

3. Take the slides through the following steps, all carried out at room temperature:

   a. Dewax in two washes of histoclear for 5 min each, followed by washes with 100%, 70%, then 50% ethanol for 5 min each, and finally wash in 0.1 *M* PBS for 5 min.

   b. Fix with 4% paraformaldehyde in 0.1 *M* PBS for 10 min (*see* **Note 6**).

   c. Wash in 0.1 *M* PBS for 5 min.

   d. Acetylate with ethanolamine solution for 10 min (to reduce nonspecific binding of the negative probe to the positively charged glass slides and tissue; *see* **Note 6**).

   e. Wash in 0.1 *M* PBS for 5 min.

   f. Dab around sections to dry. Draw around each section with a wax pen to retain applied solutions. Add enough proteinase K solution (5 μg/mL) to cover the section entirely (approx 300 μL/rat brain section), and incubate at 37°C for 5–30 min (*see* **Note 6**).

   g. Wash twice in 0.1 *M* PBS/glycine solution for 3 min each, to stop proteinase K digestion.

   h. Finally, wash in 0.1 *M* PBS for 3 min.

4. Individually tip off excess PBS, dry around sections and place slides in the Hybaid OmniSlide Thermal Cycler (*see* **Note 7**).

5. Hybridize immediately. (We have not found it necessary to prehybridize with hybridization solution minus probe before the hybridization step.)

## 3.4. Hybridization

1. Calculate the volume required to give approximately 200 ng of labeled riboprobe/ mL of hybridization buffer (*see* **Note 8**).

2. Thoroughly mix 2X hybridization solution with an equal volume of deionized formamide (*see* **Note 9**) to give the required volume of hybridization buffer (i.e., 1X hybridization solution, containing 50% [v/v] formamide; the final solution should be of pH 7.0–7.5). Add ssDNA to give a final concentration of 100 μg ssDNA/mL hybridization mixture.

3. Add labeled probe to the hybridization buffer, mix well, and spin down to remove air bubbles. Heat to 60–70°C for 10 min to denature the probe, and then cool immediately on ice for 2–3 min, to keep the probe single-stranded.

4. Pipet the determined aliquots of this hybridization buffer onto each section, and, using forceps, gently cover with a piece of Parafilm cut to the size of the section to prevent dehydration. It is important to cover the entire section with buffer, without scoring the section with the pipet tip or creating air bubbles, which are easily produced by excess pipeting due to the BSA in the solution, or by dropping the coverslip over the section.

5. Seal the hybridization chamber, and incubate overnight at the hybridization temperature. We successfully hybridize the Y-receptor riboprobes at 45°C (*see* **Note 10**).

## 3.5. Posthybridization Washing

The temperature, wash durations, and SSC concentrations used at this stage depend on the properties of the specific riboprobes used (*see* **Note 11**). The conditions described below work well for the approximately 200–600 base-long Y-receptor riboprobes and can be used as a guide for similar probes:

1. Dilute 20X SSC to the dilutions required below, and equilibrate these wash solutions to the necessary wash temperatures (takes approximately 30 min in the Hybaid wash modules; see **Note 7**). RNase-free conditions do not need to be maintained at this stage (*see* **Note 1**). Prepare enough solution to immerse the slides completely, as it is important that the sections do not dry out at any stage after hybridization, which will result in high background.
2. After hybridization, stack the slides into slide racks, and place them in slide boxes containing 2X SSC at room temperature. Wash the slides for 5 min with slight agitation, and, using forceps, carefully remove the Parafilm coverslips, which should easily float away from the sections, leaving them undamaged.
3. Once all the coverslips have been teased away from the sections, wash the slides as follows: 0.2X SSC at 55°C for 30 min, 0.1X SSC at 60°C for 30 min, followed by 25 g/mL RNaseA solution at 37°C for 15 min (*see* **Note 12**).
4. Remove residual RNase in 2X SSC at room temperature for 5 min, and finally wash in 0.2X SSC at 37°C for 30 min.

## 3.6. Probe Detection

1. Wash the slides twice in 1X TBS for 10 min each.
2. Individually dry around each section, and circle with a wax pen to retain the subsequently applied solutions. Place slide horizontally in a rack, and immediately cover each section with blocking buffer.
3. Repeat **step 2** for every slide, making sure there is no delay between removing one solution and adding the next to avoid dehydration. This is important to minimize nonspecific background signal.
4. Incubate the sections overlaid with blocking buffer for 30 min at room temperature in a humid environment.
5. Tip off the blocking buffer, and immediately add antibody solution, treating each slide separately, as in **step 3**, to avoid possible dehydration.
6. Wash slides three times in fresh 1X TBS for 5 min each.
7. Incubate slides in alkaline phosphatase buffer for 10 min.
8. Individually dry around each section, reapply wax if necessary, then place horizontally in a slide rack, and immediately apply color substrate solution to each section.
9. Repeat **step 8** for every slide, again making sure there is no delay between removing one solution and adding the next, to avoid dehydration,
10. Incubate the sections overlaid with color substrate solution overnight at room temperature in a dark and humid environment (*see* **Note 13**).

11. Wash the slides in 1X TE for 10 min to stop the reaction. Then rinse several times in fresh distilled $H_2O$, and mount sections under coverslips using Aquamount mounting medium.

## 3.7. Cellular Analysis

1. Cells are considered positively labeled under the light microscope at ×400 magnification if they show a halo pattern of brown/blue precipitate around a distinct nucleus (*see* **Figs. 3** and **4**).
2. Positively labeled cells are examined and photographed under brightfield microscopy and the results are interpreted with respect to results obtained from control experiments (*see* **Subheading 3.8.** and **Note 14**).

## 3.8. Controls

### 3.8.1. Negative Controls

Antisense versus sense probes: replace the antisense probe with a labeled sense probe, which has a complementary sequence to that of the antisense (i.e., an identical sequence to the mRNA under investigation) and therefore will have similar physical properties to the antisense probe but should not hybridize under identical assay conditions (*see* **Note 3**, **Fig. 3A–D** vs **E**, and **Fig. 4a.i** vs **b.i**; *see also* **Note 15**).

### 3.8.2. Positive Controls

1. If possible, it is useful to test the ISHH protocol on cell lines highly expressing the mRNA of interest (*see* **Fig. 4**).
2. Repeat experiments using another probe designed to a different region of the same mRNA to reconfirm results (*see* **Note 3** for design of probes), for example, in localization of rat Y4 receptor (*see* **Fig. 2** and **Note 15**).

## 4. Notes

1. It is critical to maintain RNase-free conditions prior to and during hybridization. To minimize RNase contamination, bake glassware overnight at 200°C and autoclave pipet tips, microcentrifuge tubes, and solutions when possible. Items that cannot be autoclaved or baked should be sterilized with 70% ethanol and then rinsed thoroughly with DEPC-treated distilled $H_2O$. Always wear gloves and avoid breathing directly on RNase-free items, as RNase is present on skin and in breath. Make solutions with RNase-free (e.g., DEPC-treated or autoclaved) distilled $H_2O$. (DEPC cannot be added directly to Tris-containing solutions, as primary amines will be produced.) RNA/RNA and DNA/RNA hybrids are RNase resistant, so non-RNase-free procedures can be carried out after hybridization.
2. RNase-free spun columns are commercially available (e.g., IBL Nuclean D25 columns [store at 4°C] or Biospin 30; Bio-Rad, NSW, Australia; #732-6004) for DNA >20 bp). However, it is cheaper and more straightforward to make them in-house, provided that RNase-free conditions can be maintained.

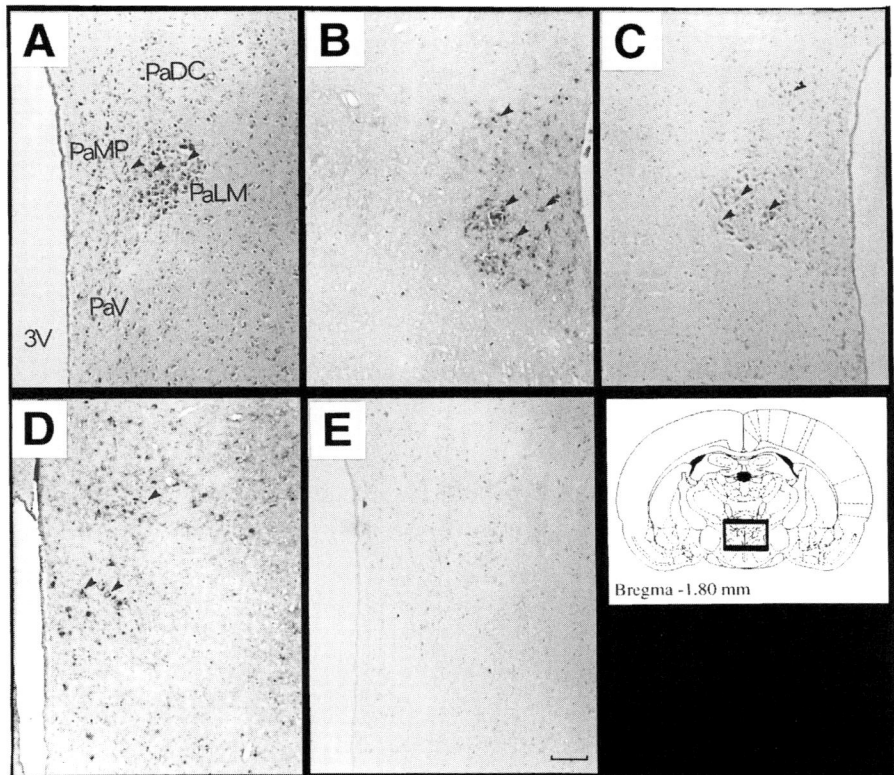

Fig. 3. Low-power photomicrographs of Y-receptor subtype mRNA expression in the paraventricular hypothalamic nucleus (PVN) of rat brain in coronal orientation. Highly specific Y-receptor subtype mRNA expression can clearly be detected (**A,** Y1; **B,** Y2; **C,** Y4; **D,** Y5) within discrete subdivisions of the PVN. No positive hybridization is detected when each of the respective sense riboprobes is used under identical assay conditions, as represented here with the $Y_1$-receptor sense probe (**E**). Scale bar = 100 μm. Arrow heads indicate positively labeled cell bodies. Boxed area in the schematic diagram (adapted from **ref. 24**) highlights the region displayed. 3V = third ventricle; PaDC = dorsal capsule of paraventricular hypothalamus; PaLM = lateral magnocellular part of paraventricular hypothalamus; PaMP = medial parvicellular part of paraventricular hypothalamus; PaV = ventral part of paraventricular hypothalamus.

3. A suitable antisense riboprobe can be 50–1000 bases long, depending on the tissue type and the way this tissue has been fixed and pretreated, all of which affect the degree of probe access to hybridization sites. When designing a suitable vector to generate a riboprobe, select from the restriction digest map an area within

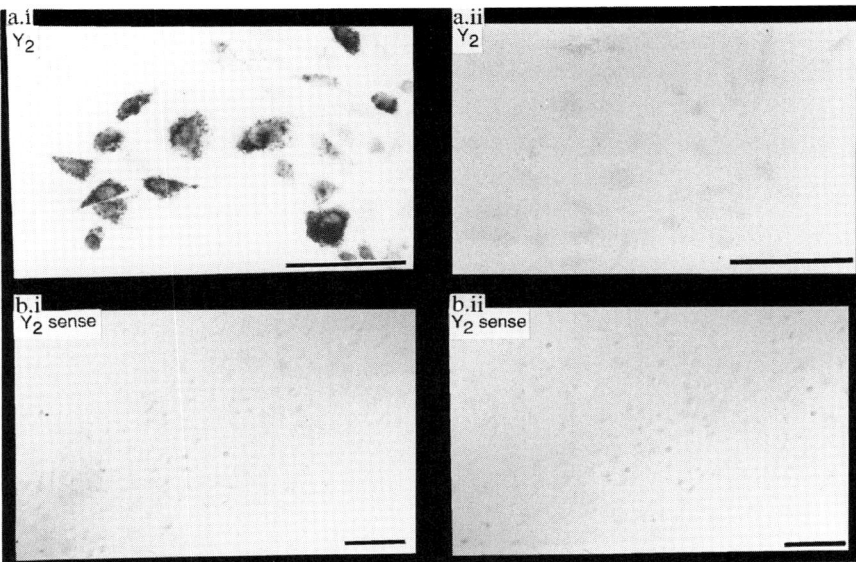

Fig. 4. Optimization of *in situ* hybridization conditions. CHO cell line controls (CHO-controls) and CHO cell lines transiently expressing human Y2 receptor (CHO-Y2) were grown in adjacent wells of 4-well chamber slides, which had been coated with poly-D-lysine/fibronectin solution for approx 2 h at room temperature. At confluence, the medium was removed, the slides were rinsed in 0.1 *M* PBS, and the cells were hybridized as described in **Subheading 3.3. steps 4b** and onwards, with either sense or antisense riboprobes generated to the human Y2 receptor. (**a.i**) Antisense on CHO-Y2 cells. (**a.ii**) Antisense on CHO-controls. (**b.i**) Sense on CHO-Y2 cell lines. (**b.ii**) Sense on CHO-controls. It is clearly demonstrated that a the positive hybridization signal is specifically obtained (halo pattern of brown/blue precipitate around a distinct nucleus) only when antisense riboprobes against CHO cell lines expressing the Y2 receptor are used. Scale bar = 100 μm.

the transcribed sequence of interest that can be subcloned into an appropriate vector expression system for in vitro transcription and that has a GC content of approximately 50%, but not so high that the probe will be very "sticky," causing background problems; the cDNA template should be specific for that sequence when compared against a database containing all other known DNA sequences, using, for example, a FASTA or BLAST search. Once this selected cDNA sequence has been subcloned, the resulting vector is amplified by a host system and then purified. The vector is cut with an appropriate restriction enzyme; the resulting linearized cDNA template is extracted with phenol/chloroform and is ready for use in in vitro transcription reactions. Check that the cDNA has been completely linearized by gel electrophoresis. An appropriate restriction enzyme is one that will cut the sequence either immediately downstream of the cDNA insert or within the insert, so that specifically sized transcripts are generated, which contain minimal nonspecific vector sequences.

Sense strands are generated by cutting at the opposite end and transcribing from the opposite promoter site. We have generated rat Y1 antisense and sense strands corresponding to coding sequence 59–654; two sets of Y4 antisense and sense strands corresponding to coding sequence 1–565 and 820–1146, respectively; a Y5 antisense strand corresponding to base sequence 461–1080 and a Y5 sense strand corresponding to base sequence 1–461 (where 1 represents the first base A of the translation initiation codon, ATG). All of these have been subcloned into pBluescript plasmid vectors (Stratagene) *(19)*. Y2-receptor subtype antisense and sense riboprobes were generated from cDNA corresponding to base sequence 838–1177 (antisense) and to base sequence 1–308 (sense) of the human Y2 receptor sequence (where 1 represents the first base A of the translation initiation codon, ATG). These were subcloned into a pcDNA3 vector (Invitrogen) *(19)*. Each cDNA sequence has 100% identity with the corresponding receptor gene sequence but exhibits <60% identity with any other known sequence. For convenience, a stock of linearized cDNA template is stored at –70°C, and, when necessary, an aliquot is diluted in RNase-free distilled $H_2O$ to the required concentration for in vitro transcription.

It is best to avoid using restriction enzymes that produce 3'-overhang ends on the cDNA template. These ends can act as promoters to initiate nonspecific or wraparound transcripts. If 3' sticky ends are unavoidable, such as when using *Pst*I, it is necessary to blunt-end the cDNA before in vitro transcription. To do so, add 5 U of T4 DNA polymerase (i.e., 0.5 µL of 10 U/µL stock) per µg of cDNA template to the reaction mixture containing transcription buffer, DTT, RNasin, and linearized DNA, as detailed in **Subheading 3.1.2., step 1**. Incubate at 22°C for 15 min, then add the remaining ingredients, and proceed with in vitro transcription exactly as described.

4. The reducing agent, DDT is added to stabilize RNasin by preventing formation of disulfide bridges.

5. RNA polymerases are very labile and should be kept on ice and returned to the freezer immediately after use. Note that promoter specificity will be lost if excess polymerase is used; at high concentrations, T7 RNA polymerase may act at the T3 promoter site and vice versa.

6. The exact steps employed in prehybridization treatment may vary depending on the nature of the tissue. The length of the paraformaldehyde fixation step is critical, and times may need to be optimized to produce sufficient tissue fixation, without causing excessive crosslinkage, which will inhibit probe penetration. The proteinase K step is also critical and needs to be optimized for each tissue block. For the conditions described here, we find an incubation time in the range of 5–10 min is usually sufficient. We routinely acetylate CNS sections, but this step may be ineffective in other situations and for other tissues.

7. We use a Hybaid OmniSlide System, but it is possible to carry out hybridization using an incubator in conjunction with sealable plastic hybridization boxes, containing thin foam or filter paper saturated with water to maintain a humid environment throughout overnight hybridization. Equally, the post-wash steps can be carried out in beakers of solution in a waterbath.

8. We successfully use 20 μL of 1X hybridization buffer/rat spinal cord section, 50 μL of 1X hybridization buffer/rat coronal brain section and 200 μL of hybridization buffer/human brain section (approximately 4 cm$^2$). The volume of hybridization buffer used will obviously depend on the size of the section. Aim to have minimum volume and therefore maximum probe concentration, yet sufficient solution that the section is completely covered and will not dry out. The optimal probe concentration should be determined empirically, by testing a range of probe concentrations above and below 200 ng/mL.

9. Formamide is a hydrogen bond breaker; therefore it acts as a destabilizing agent, reducing nonspecific binding. The signal-to-noise ratio can be manipulated by adjusting the percentage of formamide in the hybridization buffer.

10. The optimal hybridization temperature is, as a general rule, at 20–25°C below the melting temperature ($T_m$) of the probe *(21–23)*, where:

$$T_m \text{ for RNA/RNA} = 79.8 + 18.5 \times \log [Na^+] - 0.35 \times (\% \text{ formamide})/ + 58 \times (G + C) + 12 \times (G + C)^2 - 820/(\text{probe length})$$

Therefore, the $T_m$ value depends on the probe length, its base composition (GC pairs have a greater influence on overall duplex stability, as these form three hydrogen bond interactions versus AT pairs, which form two hydrogen bonds), homology to the target sequence, ionic strength of the monovalent cations present, and amount of denaturing agent (e.g., formamide) used. Hybrids formed between RNA and riboprobes are more stable than those formed with short DNA oligonucleotides; therefore the former can withstand more stringent hybridization conditions.

11. Posthybridization wash stringency is directly proportional to temperature (where the most stringent wash is recommended to be at approximately at 10–15°C below the $T_m$ value; **refs. *21–23***) and inversely proportional to the salt concentration. Therefore, the optimum ratio of signal to nonspecific background (due to weak homology with related RNA species or nonspecific interactions with other cellular components) can be manipulated by adjusting the temperature and/or the salt concentration at this stage. If these conditions are too stringent, the probe will be stripped off completely, yielding no signal detection. It is therefore vital to include positive and negative controls to test the ability to obtain consistent, sensitive, and specific signal detection.

12. RNaseA destroys single-stranded RNA, leaving duplexed RNA intact. This is therefore an important step in removing any nonhybridized riboprobe and thus reducing background. Keep all containers and solutions containing RNaseA away from everything that may come into contact with materials used prior to and during subsequent hybridization assays.

13. The incubation time required for color substrate solution will depend on the level of target RNA. Blue precipitate will form over the sections if the reaction is left too long. We add levamisole to inhibit endogenous alkaline phosphatase activity. Other procedures such as HCl bleaching may also be necessary in some tissues such as intestine, which contain high amounts of alkaline phosphatase.

14. The level of color precipitate is not linearly correlated with the amount of RNA; therefore analysis can only really be considered semiquantitative at best. Furthermore, absence of detection may only reflect lack of sensitivity of the protocol and not absence of RNA expression. It may be necessary to use *in situ* RT-PCR to detect very low-expressing-mRNA. Conversely, the presence of mRNA does not automatically demonstrate the presence of the translated peptide product.

15. Other negative controls that may be employed to analyze specific hybridization include competition studies with 100-fold excess of unlabeled antisense probe coapplied with labeled antisense probe. Other positive controls include (1) confirming the identity of the detected RNA species by molecular weight determination, using Northern hybridization; (2) assay, in conjunction with experimental tissue, of tissues known to express the signal of interest discretely; and (3) probing for a constitutive mRNA (e.g., β-actin), to verify the tissue RNA integrity.

## References

1. Gall, J. G. and Pardue, M. L. (1969) Formation and detection of RNA-DNA hybrid molecules in cytological preparations. *Proc. Natl. Acad. Sci. USA* **63,** 378–383.
2. John, H. A., Birnstiel, M. L., and Jones, K. W. (1969) RNA-DNA hybrids at the cytological level. *Nature* **223,** 582–587.
3. Höltke, H. J. and Kessler, C. (1990) Non-radioactive labeling of RNA transcripts in vitro with the hapten digoxigenin (DIG); hybridization and ELISA-based detection. *Nucleic Acids Res.* **18,** 5843–5451.
4. Schaeren-Wiemers, N. and Gerfin-Moser, A (1993) A single protocol to detect transcripts of various types and expression levels in neural tissue and cultured cells: in situ hybridisation using digoxigenin-labelled cRNA probes. *Histochemistry* **100,** 431–440.
5. Höltke, H. J., Ankenbauer, W., Mühlegger, K., Rein, R., Sagner, G., Seibl, R., et al. (1995) The digoxigenin (DIG) system for nonradioactive labelling and detection of nucleic acids—an overview. *Cell. Mol. Biol.* **41,** 883–905.
6. Panoskaltsis-Mortari A. and Bucy R. P. (1995) In situ hybridisation with digoxigenin-labelled RNA probes: facts and artifacts. *Biotechniques* **18,** 300–307.
7. Steel, J. H., Jeffery, R. E., Longcroft, J. M., Rogers, L. A., and Poulsom, R. (1998) Comparison of isotopic and nonisotopic labelling for in situ hybridisation of various mRNA targets with cRNA probes. *Eur. J. Histochem.* **42,** 143–150.
8. Blomqvist, A. G. and Herzog, H. (1997) Y-receptor subtypes—how many more? *Trends Neurosci.* **20,** 294–298.
9. Larhammar, D. (1996) Evolution of neuropeptide Y, peptide YY and pancreatic polypeptide. *Regul. Pept.* **62,** 1–11.

10. Dumont, Y., Martel, J. C., Fournier, A., St-Pierre, S., and Quirion, R. (1992) Neuropeptide Y and neuropeptide Y receptor subtypes in brain and peripheral tissues. *Prog. Neurobiol.* **38,** 125–167.
11. Heilig, M. and Widerlöv, E. (1995) Neurobiology and clinical aspects of neuropeptide Y. *Crit. Rev. Neurobiol.* **9,** 115–136.
12. Wettstein, J. G., Earley, B., and Junien, J. L. (1995) Central nervous system pharmacology of neuropeptide Y. *Pharmacol. Ther.* **65,** 397–414.
13. Eva, C., Keinänen, K., Monyer, H., Seeburg, P., and Sprengel, R. (1990) Molecular cloning of a novel G protein-coupled receptor that may belong to the neuropeptide receptor family. *FEBS Lett.* **271,** 81–84.
14. St-Pierre, J. A., Dumont, Y., Nouel, D., Herzog, H., Hamel, E., and Quirion, R. (1998) Preferential expression of the neuropeptide Y Y1 over the Y2 receptor subtype in cultured hippocampal neurons and cloning of the rat Y2 receptor. *Br. J. Pharmacol.* **123,** 183–94.
15. Lundell, I., Statnick, M. A., Johnson, D., Schober, D. A., Starbäck, P., Gehlert, D. R., et al. (1996) The cloned rat pancreatic polypeptide receptor exhibits profound differences to the orthologous receptor. *Proc. Natl. Acad. Sci. USA* **93,** 5111–5115.
16. Gerald, C., Walker, M. W., Criscione, L., Gustafson, E. L., Batzl-Hartmann, C., Smith, K. E., et al. (1996) A receptor subtype involved in neuropeptide-Y-induced food intake. *Nature* **382,** 168–171.
17. Gregor, P., Feng, Y., DeCarr, L. B., Cornfield, L. J., and McCaleb, M. L. (1996a) Molecular characterization of a second mouse pancreatic polypeptide receptor and its inactivated human homologue. *J. Biol. Chem.* **271,** 27,776–27,781.
18. Lundell, I., Statnick, M. A., Johnson, D., Schober, D. A., Starbäck, P., Gehlert, D. R., et al. (1996) The cloned rat pancreatic polypeptide receptor exhibits profound differences to the orthologous receptor. *Proc. Natl. Acad. Sci. USA* **93,** 5111–5115.
19. Parker, R. M. C. and Herzog, H. (1999) Regional distribution of Y-receptor subtype mRNAs in rat brain. *Eur. J. Neurosci.* **11,** 1431–1448.
20. Parker, R. M. C. and Herzog, H. (1998) Comparison of Y-receptor subtype expression in the rat hippocampus. *Regul. Pept.* **75–6,** 109–115.
21. Hames, B. D. and Higgins, S. J., eds. (1987) *Nucleic Acid Hybridisation: A Practical Approach.* IRL Press, Oxford University Press, Oxford.
22. Valentino, K. L., Eberwine, J. H., Barchas, J. D., eds. (1987) *In Situ Hybridisation: Applications to Neurobiology.* Oxford University Press, New York, pp. 57–58.
23. Wilkinson D. G., ed. (1993) *In situ Hybridisation: A Practical Approach.* IRL Press, Oxford University Press, New York.
24. Paxinos, G. and Watson, C. (1997) *The Rat Brain in stereotaxic coordinates*, compact 3rd ed. Academic, San Diego, CA.

# 14

## Central Y4 Receptor Distribution

*Radioactive Ribonucleotide Probe* In Situ *Hybridization with In Vitro Receptor Autoradiography*

### Philip J. Larsen and Peter Kristensen

### 1. Introduction

A large variety of different neuropeptides function as central transmitters in neuronal circuits regulating diverse aspects of energy homeostasis such as feeding behavior, energy expenditure, and thermoregulation. Neuropeptide Y (NPY) represents one such peptide mediating behavioral and homeostatic alterations favoring positive energy balance *(1)*. Thus, direct application of NPY into the hypothalamic paraventricular nucleus (PVN) or adjacent perifornical area (PeF) elicits a powerful orexigenic response leading most species to eat even in the satiated state *(2,3)*. Application of NPY to these central sites also switches the tone of the autonomic nervous system in favor of a parasympathetic tone, leading to lowered energy expenditure and insulin secretion *(4,5)*. Central administration also lowers body core temperature, but the exact site(s) and mode of action underlying this effect are largely unknown. Given the involvement of medial preoptic and hypothalamic ventromedial neurons in body temperature regulation, the numerous NPY-containing terminals within these forebrain areas are possible mediators of temperature-regulating actions of NPY.

Several studies have confirmed the involvement of an NPY-containing pathway from the ventromedial part of the arcuate (Arc) nucleus to the PVN in regulation of food intake. The level of NPY expression in this pathway is inversely correlated to circulating levels of leptin such that increasing adiposity causes NPY expression in the Arc to decrease, with resulting lowering of appetite. The receptor subtypes responsible for mediating these effects of NPY on

From: *Methods in Molecular Biology, vol. 153: Neuropeptide Y Protocols*
Edited by: A. Balasubramaniam © Humana Press Inc., Totowa, NJ

energy homeostasis are still a subject of debate. Pharmacological evidence suggests that both Y1 and Y5 receptors could be involved *(6)*, but the phenotypes resulting from knocking out expression of functional Y1 and Y5 receptors display no overt disturbances of body weight and food intake *(7,8)*. Given the redundancy of central control of energy homeostasis, this finding is not surprising, but it is of interest that treatment with Y5 antisense oligoribonucleotides induces long-lasting anorexia *(9,10)*. Also, electrophysiological studies with receptor-selective agonists suggest that Y1 and Y2 as well as Y5 receptors inhibit inhibitory postsynaptic current (IPSC)s in the parvocellular portion of the PVN *(11)*, indicative of NPY-mediated presynaptic inhibition of γ-aminobutyric acid release via multiple receptors.

The NPY receptor family has been pharmacologically defined by its ability to bind NPY, peptide YY (PYY), pancreatic polypeptide (PP), and their synthetic derivatives *(12–14)*. Both Y1 and Y5 receptors are expressed in hypothalamic nuclei centrally involved in feeding behavior *(15,16)*, but Y2 receptor mRNA is also present in several hypothalamic nuclei *(17)*. Less is known about the distribution of central Y4 receptors, but recent reports of regional Y4 mRNA in the hypothalamus and dorsal vagal complex *(18,19)* suggest that this receptor may also be involved in mediating central effects of NPY on energy homeostasis, although most pharmacological experiments have failed to show effects of rat PP on feeding behavior. The rank order of binding potency for the Y4 receptor is as follows: PP > NPY, PYY *(20–22)*.

To examine the central distribution of Y4 receptors in the rat brain, we first determined the presence and location of Y4 receptor mRNA by *in situ* hybridization histochemistry with radioactively labeled ribonucleotides. In a parallel series of experiments, receptor autoradiography with radiolabeled rat PP was carried out. We chose rat PP as the radioligand because of its high specificity and potency for Y4 receptors. By comparing *in situ* hybridization histochemical results with receptor autoradiography, it was possible to validate that expressed Y4-encoding mRNA is translated into receptor protein with full binding capacity.

## 2. Materials

### 2.1. Tissue and Sections

Adult male Wistar rats weighing approx 200 g housed under standard laboratory conditions with free access to food and water were used. Animals were decapitated early in the light phase (L/D, 12:12 h), and the brains were rapidly removed from the skull and kept frozen at –80°C until 12-μm cryostat sections were cut. Both coronal and frontal sections were cut. One-in-seven series of sections were thaw-mounted on SuperFrost Plus microscopic slides and kept at

–80°C until use. One of these series was counterstained with cresyl violet or thionin, serving as an anatomical reference.

## 2.2. In Situ *Hybridization Histochemistry*

1. Diethylpyrocarbonate (DEPC) water: add 0.1% (vol) to deionized water in the fume hood. Leave at room temperature overnight, and autoclave to remove any DEPC left.
2. Phosphate-buffered saline (PBS) 10X: 1.3 $M$ NaCl, 70 m$M$ Na$_2$HPO$_4$, 30 m$M$ NaH$_2$PO$_4$, pH 7.0, in DEPC water. Autoclave.
3. 4% Paraformaldehyde: prepared in PBS with DEPC water.
4. Labeled RNA-probe: prepare using standard procedure (as described by Promega, Stratagene, Amersham, or other commercial suppliers) for RNA transcription using SP6, T7, or T3 polymerases with [$^{35}$S]UTP. Following removal of nonincorporated nucleotide, dissolve in 10 m$M$ dithiothreitol (DTT), and add equal volume of deionized formamide before storage at –20°C until use (*see* also below).
5. t-RNA: prepare 10 mg/mL solution in DEPC-treated water. Filter-sterilize, and store at –20°C.
6. Triethanolamine: prepare a 2 $M$ solution in DEPC-treated water.
7. SALTS 10X: mix the following ingredients: 3 $M$ NaCl (35 g). 0.1 $M$ Tris (2.4 g), 0.1 $M$ Na$_2$HPO$_4$, pH 6.8, (2.84 g), 50 m$M$ EDTA (20 mL 0.5 $M$), Ficoll 400 (Sigma #F4375, St.Louis, MO; 0.2% w/v; 0.4 g), polyvinylpyrolidone (Sigma #PVP-40, 40000 MW; 0.2% w/v; 0.4 g), BSA Fraction V (Sigma #A4503; 0.2% w/v; 0.4 g), and make up to 200 mL in DEPC-treated water. Filter-sterilize, and store in small (for preparation of hybridization solution) and large (for washing) aliquots at –20°C.
8. 50% dextran sulfate (Sigma #D8906). Make up in DEPC water by stirring while heating. Keep frozen in small portions at -20°C.
9. Deionized formamide (for hybridization solution): mix 100 mL of formamide with 10 g Bio-Rad #AG501-X8 (20–50 Mesh) Mixed Bed Resin. Stir at 4°C for 30 min. Filter through Whatman filter paper. Repeat resin addition and stirring. Filter twice, and store at –20°C. For washings, use good-quality formamide straight from the bottle.
10. DTT stock: make up 1 $M$ stock with DEPC water. Filter-sterilize, and store at –20°C.
11. 1X NTE: 0.5 $M$ NaCl, 10 m$M$ Tris-HCl, pH 7.2, 1 m$M$ EDTA. Make 5X stock with deionized water and autoclave.
12. RNase A (10 mg/mL), in deionized water. Keep stock at –20°C. Use only in designated glassware kept separately from other glassware used in the laboratory.
13. Standard saline citrate (SSC) 20X: 3.0 $M$ NaCl, 0.3 $M$ trisodium citrate trinatriumcitrat, 2H$_2$O, pH 7.0, in deionized water.
14. Hybridization buffer, pH 6.8: deionized formamide 50%, dextran sulfate 10%, tRNA 1 mg/mL, Ficoll 400 0.02% v/w, Polyvinylpyrrolidone 0.02% v/w, BSA Fraction V 0.2% v/w, DTT 10 m$M$, NaCl, 0.3 m$M$, EDTA 0.5 m$M$, Tris-Cl 10 m$M$, Na$_2$HPO$_4$ 10 m$M$.

## 2.3. Equipment

Baked glassware, sterile forceps, sterile gloves, box of coverslips reserved for hybridization, fume hood and pipets reserved for RNA work.

## 2.4. In Vitro *Receptor Autoradiography*

1. Synthetic PP and PYY were purchased from Bachem (UK). All other reagents were purchased from Sigma. High-performance liquid chromatography (HPLC) purified radioiodinated [$^{125}$I]ratPP having a specific activity of approx 2000 Ci/mmol was kindly prepared by Dr. Christian Foged, Novo Nordisk A/S. *Buffers (prepare fresh just prior to use):*
2. HEPES buffer, pH 7.4: 25 m$M$ HEPES, 2.5 m$M$ CaCl$_2$, 1 m$M$ MgCl$_2$.
3. Incubation buffer, pH 7.4: 25 m$M$ HEPES, 2.5 m$M$ CaCl$_2$, 1 m$M$ MgCl$_2$, BSA 0.5 g/L, bacitracin 0.5 mg/mL and BSA 0.5 g/L, 1 m$M$ phenylmethyl-sulfonyl flouride (PMSF).
    PMSF stock solution (2M) is prepared in 96% ethanol. Add 500 µL of this to 1 L of HEPES buffer to get a final concentration of 1 m$M$ .

## 3. Methods

## 3.1. Y4 In Situ *Hybridization Histochemistry*

### 3.1.1. Cloning

Rat Y4 receptor cDNAs were cloned by polymerase chain reaction (PCR) using the primers based on the published sequences (GenBank accession number Z68180 (rY4)). PCR was carried out for 30 cycles of 1 min at 94°C, 1 min at 55°C, and 2 min at 72°C using rat genomic DNA as a template and sequence-specific oligonucleotides as primers.

### 3.1.2. Riboprobe Synthesis

Riboprobes were prepared from approx 500 bp nonoverlapping template fragments derived from the 5'-end or 3'-end of the receptor cDNA cloned into Bluescript vectors (Stratagene, La Jolla, CA). Plasmid DNA was prepared and, $^{35}$S-labeled antisense and sense RNA was transcribed using T3 or T7 polymerases as described previously *(10)*. Briefly, following transcription the probe preparations were precipitated repeatedly with ethanol after addition of ammonium acetate to a final concentration of 3.0 $M$ and hydrolyzed to an average size of 100 bp. The amount of trichloroacetic acid (TCA)-precipitable material in the final probe preparation was usually >90%. The two corresponding RNA probes transcribed from the opposite strands of the same plasmid template were adjusted to the same amount of radioactivity concentration, and all experiments included sections hybridized using the sense RNA probes.

### 3.1.3. Pretreatment

1. Thaw sections at room temperature for 10 min. Fix in 4% paraformaldehyde (PFA) for 5 min. Wash briefly twice in PBS, leave in PBS for 5 min twice, and wash for 2 min in PBS.

   In some tissues pretreatment with proteinase K is carried out prior to hybridization. This step improves penetration of ribonucleotides in tissues with high connective tissue content. Given the low amount of such extracellular matrix in the central nervous system (CNS), this treatment does more harm to the overall morphology than good to the sensitivity of the hybridization signal in sections from the brain (*see* **Note 1**).
2. Acetylation (in fume hood): 2 min in 0.1 $M$ triethanolamine (from 2 $M$ stock with DEPC water). While stirring add 0.25 vol% acetic anhydride, and wait for 2 min.
3. Wash for 2 min in PBS.
4. Dehydrate sections using a series of EtOH (30%, 60%, 80%, 96%, 99%, 99%) treatments for 10 s each. Leave to air-dry before hybridization.

### 3.1.4. Hybridization

1. Hybridization is performed at 47°C overnight in 50% formamide containing 10 m$M$ DTT, placed in a closed humid chamber.
2. Hybridization solution (100 µL/section) is made up using probe solution (20 µL), 10X SALTS 10 µL), Deionized formamide (40 µL), 50% dextran sulfate (10 µL, t-RNA 9.2 µL, and 1.0 $M$ DTT 0.8 µL. Before addition, the probe solution is denatured at 80°C for 3 min.
3. Probe mix is added to the dry sections, and the section is covered by a clean coverslip.

### 3.1.5. Posthybridization Washing

All steps are carried out with gentle agitation except the first two wash and dehydration steps.

1. Wash for 1 h at 52/57° or 62°C in 50% formamide/1X SALTS/10 m$M$ DTT followed by 1 hour at 57/62 or 67°C in 50% formamide/1X salts/10 m$M$ DTT. Carry out washes in a water bath to maintain appropriate temperature.
2. Rinse sections for 5 min at room temperature in prewarmed (44°C) NTE with 10 m$M$ DTT.
3. Remove single-stranded RNA by treatment with RNaseA. Incubate for 30 min in NTE with 20 µg/mL RNaseA, taking great care not to contaminate anything in the lab with RNase.
4. Rinse sections $2 \times 5$ min at room temperature in NTE with 10 m$M$ DTT.
5. Wash for 30 min at room temperature in 0.1X SSC/1 m$M$ DTT.
6. Dehydrate through series of EtOH (30%, 60%, 80%, 90%, $2 \times 99\%$) containing 0.3 $M$ NH$_4$OAc.
7. After exposure to β-radiation-sensitive films (Hyperfilm, Amersham, Birkerød, Denmark), sections were dipped in Amersham LM-1 nuclear emulsion and exposed for 9 days before developing in Kodak D19. Emulsion-dipped sections

were Nissl counterstained and topographic analysis of the microscopic localization of Y4 mRNA expressing cells was facilitated by using a microscope fitted with a camera lucida device. Background labeling was evaluated over areas dominated by white matter, and cells were considered positively labeled when the density of overlying photographic silver grains was higher than five times the background labeling (*see* **Note 1**). At representative rostrocaudal levels of the SCN, numbers of positively hybridized cells were counted and compared with the number of Nissl-stained neurons. The approximate percentages of positively hybridized neurons were subsequently estimated.

### 3.1.6. [$^{125}$I]RatPP In Vitro Receptor Autoradiography

Receptor localization and characterization has traditionally been carried out on isolated cell membrane preparations obtained from tissues suspected to contain specific binding sites for hormones, neurotransmitters, growth factors, and so forth. However, the very heterogeneous topography of the CNS does not allow such simple procedures if the exact localization of the binding site is needed (*see* **Note 2**), rather it is necessary to visualize the binding sites *in situ* on sections or cultures from specific areas of the CNS. Visualization of receptors in specific regions of the CNS can be obtained by incubating unfixed tissue sections with a radioactively labeled ligand; the subsequent photographic detection of the radioisotope in the sections is known as receptor autoradiography (in vitro autoradiography). In the present studies, an in vitro autoradiographic method has been developed to visualize rat PP binding, which was labeled with the radioisotope $^{125}$I, a β-radiation emitter as well a photoelectric radiation emitter (due to internal conversion of electrons). The resolution obtained by autoradiography is regional if sections are exposed to X-ray films or special β-radiation-sensitive films, whereas a resolution at the light or electron microscopic level is obtained if the sections are emulsion dipped after binding of the ligand (*see* **Note 2**). A successful outcome of the latter technique depends to a large extent on the solubility of the ligand since diffusion from the "true" binding site should be minimized. In the present study, binding experiments were used to verify expression of the Y4 receptor protein, and consequently no further attempts were made to visualize cellular labeling.

1. Take sections out of the freezer. Keep them on dry ice until use.
2. Transfer sections to a hot plate (40°C) for 1 min to remove condensed water. Transfer the sections to a Couplin jar containing the preincubation buffer. Ensure that the sections are properly covered with the incubation buffer.
3. Preincubate in binding buffer at room temperature for 20 min, after which the glass slides are lifted from the jar and left vertically for 1 min on a piece of tissue paper to allow drainage of excess incubation buffer.
4. Incubate with [$^{125}$I]PYY at room temperature for 60 min in the incubation buffer to which the radioligand is added (0.1 n*M* , total binding). A full series of neigh-

boring sections are incubated with incubation buffer to which 0.1 n$M$ [$^{125}$I]ratPP and excess unlabeled NPY/PYY are added (6 μ$M$ , nonspecific binding).

Note that sections are preincubated and washed in Couplin jars, while the incubation with the radioligand takes place in a humid incubation chamber consisting of a flat Perspex container with a loose fitting lid. Tissue paper is placed in the bottom of the chamber and soaked in water. Aluminum bars to carry the sections are placed on top of the tissue paper. Slides are placed horizontally on the aluminum bars and covered by 200 μL of incubation buffer to which the radioligand is added. It is most important to ensure that the incubation buffer covers all of the sections on the slide. After incubation with the radioligand, the slides are removed from the incubation chamber, allowed to drain excess incubation buffer, and transferred into a series of washes.

5. Wash briefly in incubation buffer at 4°C for 10 s. Washes are carried out in a line of Couplin jars placed in tray full of crushed ice.
6. Wash 4 × 10 min in incubation buffer at 4°C.

   Peptides of the PP family are notorious for their "stickiness," i.e., high degree of nonspecific binding, particularly to white matter. However, the presently employed [$^{125}$I]PP ligand does not represent a problem concerning high levels of background labeling, whereas [$^{125}$I]NPY represents a far larger problem, with nonspecific binding often being >70% of the total binding (*see* **Note 3**). To reduce this nonspecific binding, 0.01% of the detergent Triton X-100 can be added to the washing buffer, a procedure that considerably reduces levels of background binding.
7. Wash briefly in distilled water at 4°C. Thereafter, the slide is placed vertically on a thick piece of absorbant tissue paper while it is blow-dried under a gentle stream of cool air. It is absolutely essential that this last step be carried out with speed and efficiency to reduce the diffusion of the radioligand as much as possible.
8. Slides are mounted on a piece of card in a light-proof box, exposed to an X-ray film (HyperfilmTM, Amersham) for 1 week, and finally developed in Kodak D19 developer. Autoradiograms are quantified using an image analysis system (Image1.59, courtesy of Wayne Rasband, NIH), and grain densities are converted to fmol-bound ligand/mg wet weight using either $^{35}$S (*in situ* hybridization) or $^{125}$I (receptor autoradiography) microscales as reference standards (Amersham). The measured grain densities over individual areas can be corrected for background nonspecific binding by subtracting grain density values of the same area from neighboring sections, which are incubated under conditions visualizing nonspecific binding, i.e., containing a huge surplus of unlabeled ligand.

## 4. Notes

1. The specificity of *in situ* hybridization procedures is verified by testing both sense and antisense probes. In the present study, the Y4 sense probe gave rise to labeling only in the cornua ammonis (**Fig. 1**), where many sense probes consistently gave less intense hybridization signals. This hippocampal nonspecific hybridization may have something to do with a particularly high cell density in the dentate

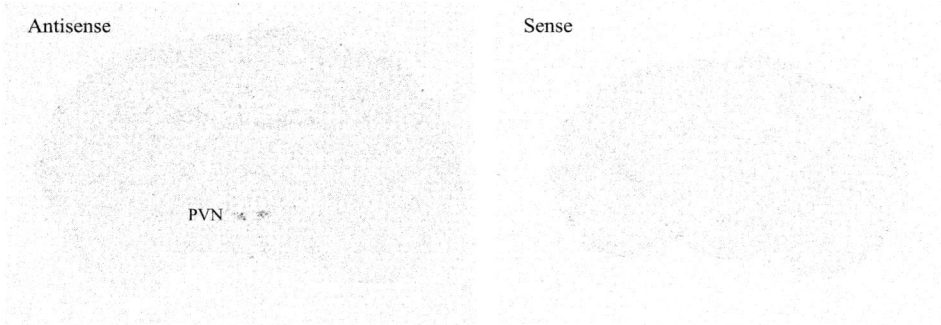

Fig. 1. Autoradiograms of neighboring sections showing *in situ* hybridization histochemical reaction for Y4 mRNA in the forebrain at the level of the central portion of
the hypothalamic paraventricular nucleus. As clearly seen, specific Y4 expression is
detected in the ventral portion of the medial parvicellular subnucleus of the PVN. The
specificity of the hippocampal signal is uncertain as the density of the overlying silver
grains is identical for both antisense and sense probe hybridization.

gyrus because it is often present also in the granular layer of the cerebellar cor-
tex. Specificity of the hybridization signal was further evaluated in emulsion-
dipped slides, and the signal was considered positive when the density of silver
grains was five times higher than background. Assuming that slides are developed with the sensitivity of the emulsion still in the linear range, and consequently follow a Poisson distribution, this threshold ensures false-positive
observations with a probability <2.5%. Currently, positive Y4 hybridization signals have been detected in very few brain regions, with the hypothalamic
paraventricular nucleus being the only forebrain area with specific Y4 mRNA
expression. At the central level of the PVN (corresponding to plate 26 in **ref. 23**),
low to moderate densities of silver grains were seen over cells confined to the
medial, lateral, and periventricular parvicellular subnuclei. The highest density
of Y4 mRNA expression in the PVN was observed in the ventral portion of the
medial parvicellular part of the PVN. Other hypothalamic nuclei were completely
devoid of Y4 mRNA (**Fig. 2**).

Of particular interest is the complete lack of Y4 expression in the supraoptic
nucleus, suggesting that none of the positively labeled cells in the PVN are
magnocellular neurons projecting to the neurohypophysis. In the mesencepha-
lon, a positive hybridization signal of equal density to that in the PVN was seen
in the nucleus of the trapezoid body. In the caudal part of the brainstem, positive
Y4 hybridization signals were detected in the rat dorsal vagal complex at the
light microscopic level. High densities of silver grains were observed over neurons of the dorsal vagal motor nucleus and over neurons of a subregion of the
nucleus of the solitary tract known as the subnucleus gelatinosus. Furthermore,

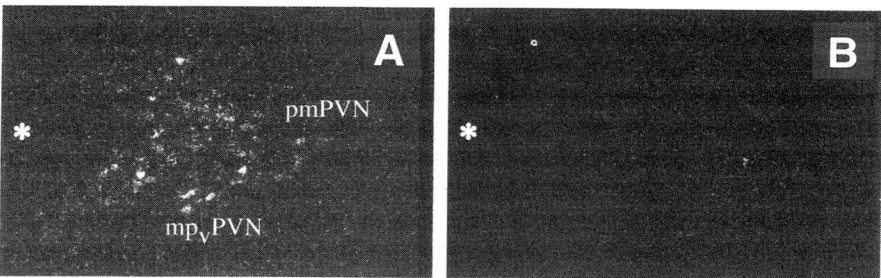

Fig. 2. Darkfield autoradiograms demonstrating Y4 receptor expression in the ventral aspect of the medial parvicellular PVN. (**A**) Antisense probe hybridization. (**B**) Sense probe hybridization.

cells within the ventral margin of the area postrema expressed high levels of Y4 mRNA (**Fig. 3**).

2. In vitro receptor autoradiography visualized central Y4 binding sites in both the PVN and the dorsal vagal complex, displaying a complete topographical overlap with Y4 mRNA distribution. Thus, high levels of displaceable [$^{125}$I]ratPP binding were seen in the PVN (**Fig. 4**). Nonspecific binding (in the presence of 1 m*M* unlabeled PYY) did not give rise to detectable labeling of tissue sections. We did not attempt to validate the character of the binding sites with further kinetic analysis, but given the restricted localization completely overlapping that of the Y4 hybridization signal it seems evident that the visualized binding sites are truly Y4 receptors. Using *in situ* hybridization histochemistry with $^{35}$S-labeled riboprobes, we have demonstrated that the central expression of Y4 receptor mRNA is restricted to very few nuclei, all of which are centrally involved in regulation of multiple homeostatic mechanisms.

Receptor autoradiographic analysis of the topographical distribution of NPY binding sites has helped little to identify NPY-responsive neurons in the hypothalamus mainly because very low densities of binding sites have been detected *(13,16,24,25)*. Therefore, it was surprising that [$^{125}$I]ratPP proved such an excellent radioligand because numerous alternatives to receptor autoradiography have been sought for NPY receptors. With the cloning of most NPY-receptor encoding genes, it became possible to visualize receptor-expressing neurons using *in situ* hybridization histochemistry. However, obvious interpretational limitations are connected to *in situ* hybridization histochemical mapping of receptor expression. First, it is uncertain to what degree the level of gene expression reflects receptor synthesis, and second, it is unknown whether *de novo* synthesized receptors are functional binding sites. Third, only perikarya are visualized by *in situ* hybridization histochemistry, leaving it impossible to evaluate the exact distribution of the functional receptor in the neuronal membrane (somatic, dendritic, axonal).

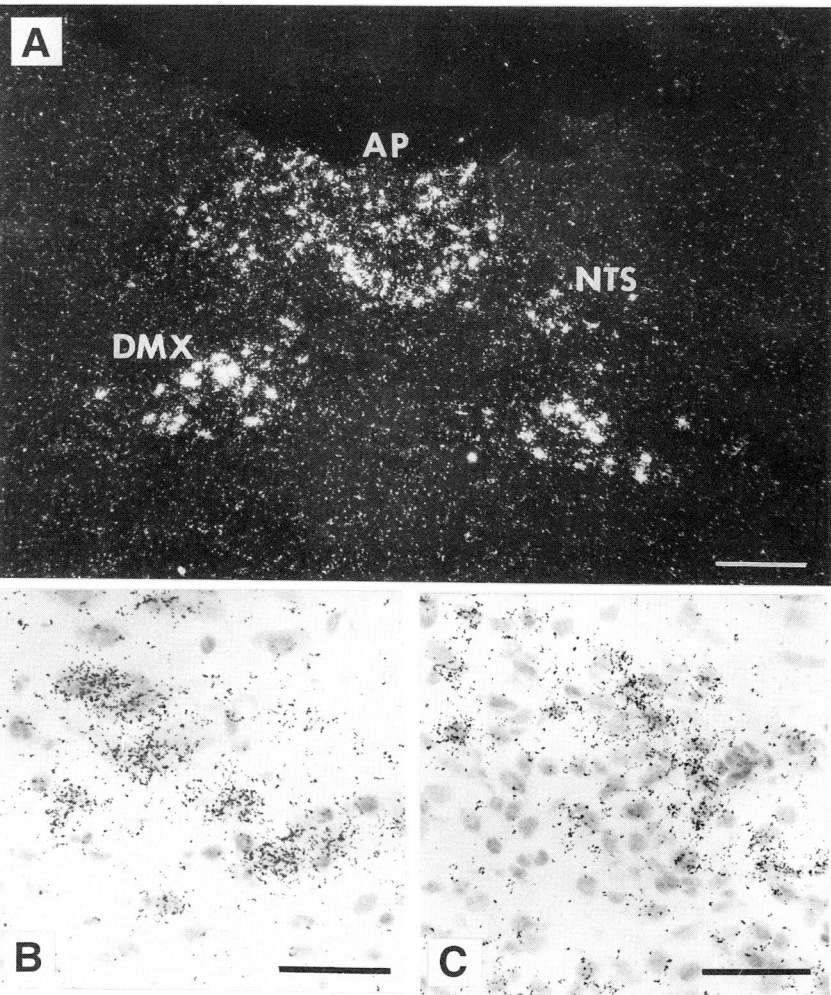

Fig. 3. (**A–C**) Emulsion-dipped slides showing Y4 mRNA in the dorsal vagal complex. AP, area postrenia; DMX, dorsal vagal motor nucleus; NTS, nucleus of the solitary tract. Scale bar-25 μm.

These limitations were easily overcome by performing a parallel series of in vitro receptor autoradiograms using rat PP as a Y4-specific ligand. Given the complete overlap of Y4 mRNA localization and rat PP binding, it is likely that most, if not all, Y4-receptor expression in the PVN gives rise to somatic binding sites. However, it is impossible to rule out completely that some of the Y4 receptors synthesized by neurons of the medial parvicellular PVN are presynaptic at

Total <sup>125</sup>I-rPP binding

Non-specific binding

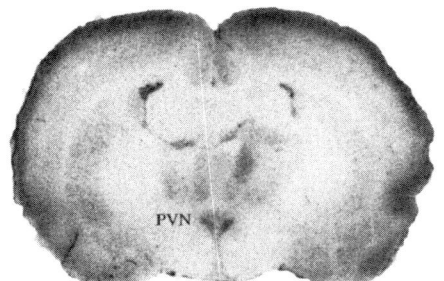 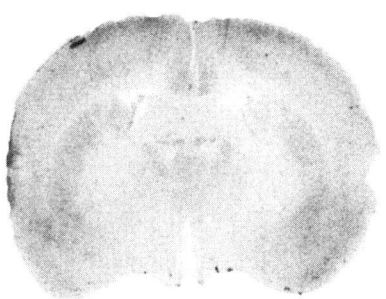

Fig. 4. In vitro receptor autoradiograms showing total [$^{125}$I]ratPP binding and nonspecific binding to unfixed frontal rat brain sections. Nonspecific binding represents binding of the raidoligand in the presence of 1 m$M$ unlabeled rat PP. The highest density of specific binding sites is seen in the PVN, but specific binding to cortical areas and thalamic nuclei is also visible.

the level of axonal terminals. The periventricular and dorsal subset of the medial parvicellular subnuclei of the PVN contains neuroendocrine cells releasing their content of hypophysiotrophic factors at the level of the external zone of the median eminence. Positive hybridization was never observed in these two subnuclei of the PVN and the median eminence displayed no [$^{125}$I]ratPP binding, excluding direct Y4-mediated actions on neuroendocrine cells in the PVN. High levels of both Y4 mRNA expression and rat PP binding were seen over neurons in the ventral portion of the medial parvicellular PVN. This subregion of the PVN projects mainly to preganglionic autonomic neurons of the dorsal vagal complex and the intermediolateral column of the thoracic and upper lumber spinal cord *(26)*, suggesting that central Y4 receptor activation may influence autonomic output.

3. The stochastic nature of radioactive decay makes quantification of the hybridization signal obtained with radiolabeled riboprobes reliable as long as detection is within the linear range of the emulsion. This holds true for both X-ray film and fluid emulsions. However, using radioactively labeled riboprobes makes it difficult to assess the morphological characteristics of labeled cells with certainty. In the CNS, the sheer size of neuronal perikarya makes confusion with glial cells unlikely, but more detailed pheonotypical characteristics are unavailable without dual immunocytochemistry/*in situ* hybridization. If levels of mRNA are sufficiently high, nonradioactive *in situ* hybridization may help solve this problem because most chromogens yield better cellular resolution than silver grains of a photographic emulsion. For high-abundance mRNA targets, the sensitivity of each method appears similar; therefore the choice between isotopic and nonisotopic labels is dependent on the aim of the study and the cellular resolution

required *(27)*. However, in our experience as well as that of others, low-abundance targets such as receptor-encoding mRNA are more readily detected by radioactively labeled riboprobes. Given the stickiness of RNA, it is essential to distinguish nonspecific adherence of riboprobes to tissue components other than RNA from true hybridization signal.

However trivial this may seem, we have tried to validate the specificity of the hybridization signal by using several probes recognizing different regions of the Y4 mRNA as well as complementary sense strands. Because of the complete absence of hybridization signal in the PVN and dorsal vagal complex with sense probes, the signals here were considered specific. However, uniform labeling of the hippocampus was seen with both antisense and sense strand hybridization with an equal tendency to disappear with increasing washing temperatures. Thus, the hybrids formed with either of these probes appear to be similarly base paired and must resist on more or less identical stringent environments, suggesting that the signals are nonspecific. Nonspecific interaction probably due to interaction with nonnucleic acid sequences is actually often seen for the hippocampus and the cerebellum, probably because of high cellular density in both of these regions. Other strategies including competition of the hybridization sites with excess cold probe or detailed melting point analysis of the hybrid thermal stability have not been carried out.

## Acknowledgments

The expert technical assistance of J. Mandelbaum is gratefully acknowledged. We also wish to thank Dr. J.S. Rasmussen (Department of Molecular Genetics, Novo Nordisk) for cloning the rat Y4 receptor cDNA.

## References

1. Turton, M. D., O'Shea, D., and Bloom, S. R. (1997) Central effects of neuropeptide Y with emphasis on its role in obesity and diabetes, in *Neuropeptide Y and Drug Development* (Grundemar, L., and Bloom, S. R., eds.), Academic, San Diego, pp. 15–39.
2. Stanley, B. G. and Leibowitz, S. F. (1985) Neuropeptide Y injected in the paraventricular hypothalamus: a powerful stimulant of feeding behavior. *Proc. Natl. Acad. Sci. USA* **82,** 3940–3943.
3. Stanley, B. G. and Thomas, W. J. (1993) Feeding responses to perifornical hypothalamic injection of neuropeptide Y in relation to circadian rhythms of eating bahavior. *Peptides* **14,** 475–481.
4. Dunbar, J. C., Hu, Y., and Lu, H. (1997) Intracerebroventricular leptin increases lumbar and renal sympathetic nerve activity and blood pressure in normal rats. *Diabetes* **46,** 2040–2043.
5. Moltz, J. H. and McDonald, J. K. (1985) Neuropeptide Y: direct and indirect action on insulin secretion in the rat. *Peptides* **6,** 1155–1159.
6. Woldbye, D. P. D. and Larsen, P. J. (1998) The how and Y of eating. *Nature Med.* **4,** 671–672.

7. Pedrazzini, T., Seydoux, J., Künstner, P., Aubert, J.-F., Grouzmann, E., Beermann, F., et al. (1998) Cardiovascular response, feeding behavior and locomotor activity in mice lacking the NPY Y1 receptor. *Nature Med.* **4,** 722–726.
8. Marsh, D. J., Hollopeter, G., Kafer, K. E., and Palmiter, R. D. (1998) Role of the Y5 neuropeptide Y receptor in feeding and obesity. *Nature Med.* **4,** 718–722.
9. Schaffhauser, A. O., Stricker-Krongrad, A., Brunner, L., Cumin, F., Gerald, C., Whitebread, S., et al. (1997) Inhibition of food intake by neuropeptide Y Y5 receptor antisense oligodeoxynucleotides. *Diabetes* **46,** 1792-1798.
10. Tang-Christensen, M., Kristensen, P., Stidsen, C. E., Brand, C. L., and P. J., L. (1998) Central administration of Y5 receptor antisense decreases spontaneous food intake and attenuates feeding in response to exogenous neuropeptide Y. *J. Endocrinol.* **159,** 307-312.
11. Pronchuk, N. and Colmers, W. F. (1998) NPY Y1, $Y_2$ and $Y_5$ receptors inhibit IPSCs in hypothalamic parvocellular PVN neurons. *Soc. Neurosci. Abstr.* **24,** 819. 4.
12. Chen, G. and van Den Pol, A., N, (1996) Multiple NPY receptors coexist in pre and postsynaptic sites: inhibition of GABA release in isolated self-innervating SCN neurons. *J. Neurosci.* **16,** 7711–7724.
13. Gehlert, D. R. (1994) Subtypes of receptors for neuropeptide Y: implications for the targeting of therapeutics. *Life Sci.* **55,** 551–562.
14. Wahlestedt, C. and Reis, D. J. (1993) Neuropeptide Y-related peptides and their receptors—are the receptors potential therapeutic drug targets? *Annu. Rev. Pharmacol. Toxicol.* **33,** 309–352.
15. Gerald, C., Walker, M. W., Criscione, L., Gustafson, E. L., Batzl-Hartmann, C., Linemeyer, D. L., et al. (1996) A receptor subtype involved in neuropeptide Y-induced food intake. *Nature* **382,** 168–171.
16. Larsen, P. J., Sheikh, S. P., Schwartz, T. W., Rehfeld-Jacobsen, C., and Mikkelsen, J. D. (1993) Regional distribution of putative NPY Y1 receptors in forebrain areas of the rat central nervous system. *Eur. J. Neurosci.* **5,** 1622–1637.
17. Gustafson, E. L., Smith, K. E., Durkin, M. M., Walker, M. W., Gerald, C., Weinshank, R., and Branchek, T. A. (1997) Distribution of the neuropeptide Y Y2 receptor mRNA in rat central nervous system. *Mol. Brain Res.* **46,** 223–235.
18. Larsen, P. J. and Kristensen, P. (1997) The neuropeptide Y (Y4) receptor is highly expressed in neurones of the rat dorsal vagal complex. *Brain Res. Mol. Brain Res.* **48,** 1–6.
19. Larsen, P. J. and Kristensen, P. (1998) Distribution of neuropeptide Y receptor expression in the rat suprachiasmatic nucleus. *Brain Res. Mol. Brain Res.* **60,** 69–76.
20. Bard, J. A., Walker, M. W., Branchek, T. A., and Weinshank, R. L. (1995) Cloning and functional expression of a human Y4 subtype receptor for pancreatic polypeptide, neuropeptide Y, and peptide YY. *J. Biol. Chem.* **270,** 26,762–26,765.
21. Lundell, I., Blomqvist, A. G., Berglund, M. M., Schober, D. A., Johnson, D., Statnick, M. A., et al. (1995) Cloning of a human receptor of the NPY receptor family with high affinity for pancreatic polypeptide and peptide YY. *J. Biol. Chem.* **270,** 29,123–29,128.

22. Yan, H., Yang, J., Marasco, J., Yamaguchi, K., Brenner, S., Collins, F., et al. (1996) Cloning and functional expression of cDNAs encoding human and rat pancreatic polypeptide receptors. *Proc. Natl. Acad. Sci. USA* **93,** 4661–4665.
23. Swanson, L. W. (1992) *Brain Maps: Structure of the Rat Brain.* Elsevier, New York.
24. Dumont, Y., Fournier, A., St-Pierre, S., and Quirion, R. (1993) Comprarative characterization and autoradiographic distribution of neuropeptide Y receptor subtypes in the rat brain. *J. Neurosci.* **13,** 73–86.
25. Dumont, Y., Martel, J. C., Fournier, A., St-Pierre, S., and Quirion, R. (1992) Neuropeptide Y and neuropeptide Y receptor subtypes in brain and peripheral tissues. *Prog. Neurobiol.* **38,** 125–167.
26. Swanson, L. W. (1987) The hypothalamus., in *Handbook of Chemical Neuroanatomy*, vol. 5 (Björklund, A., Hökfelt, T., and Swanson, L. W., eds.), Elsevier, Amsterdam, pp. 1–124.
27. Steel, J. H., Jeffery, R. E., Longcroft, J. M., Rogers, L. A., and Poulsom, R. (1998) Comparison of isotopic and nonisotopic labeling for in-situ hybridization of various messenger-RNA targets with cRNA probes. *Eur. J. Histochem.* **42,** 143–150.

# 15

## Combining Non-Isotopic Localization of NPY mRNA with Immunocytochemistry

**Jennifer Lachey, Jennifer Profitt, Chad Foradori, and Randy J. Seeley**

## 1. Introduction

This protocol combines *in situ* hybridization techniques with immunocytochemistry (ICC). Such a protocol generally allows you to identify cells that have both a specific species of mRNA expressed and a separate specific protein. This combination of techniques is particularly powerful for a system like neuropeptide Y (NPY). NPY has a broad distribution in the mammalian central nervous system (CNS) and probably has multiple functions. One way to understand those functions is to assess changes in neuronal activity of NPYergic cells using an antibody directed to the protein Fos. Fos is the product of the early immediate gene c-fos and is typically found in neurons after treatments that result in their activation. As such, it has become a broadly used and accepted marker for increased neuronal activity after a specific treatment.

This protocol allows one to apply any of a variety of treatments to an animal (for example, food deprivation) and then assess what parts of the CNS are activated by that treatment. Additionally, one can then determine whether the neurons that have showed increased activity also express mRNA for NPY and hence determine whether specific populations of NPY cells are changing their activity under a variety of circumstances. Using *in situ* hybridization to localize NPYergic cells has some specific advantages over using an antibody. Antibodies will not selectively stain for the cell body but will also stain for terminals where NPY has been transported. Cell bodies stain less heavily unless animals are treated with compounds like colchicine, which break down microtubules. Unfortunately, such compounds also produce profound changes in neuronal activity, making the Fos assessments uninterpretable. The current protocol gets around these

From: *Methods in Molecular Biology, vol. 153: Neuropeptide Y Protocols*
Edited by: A. Balasubramaniam © Humana Press Inc., Totowa, NJ

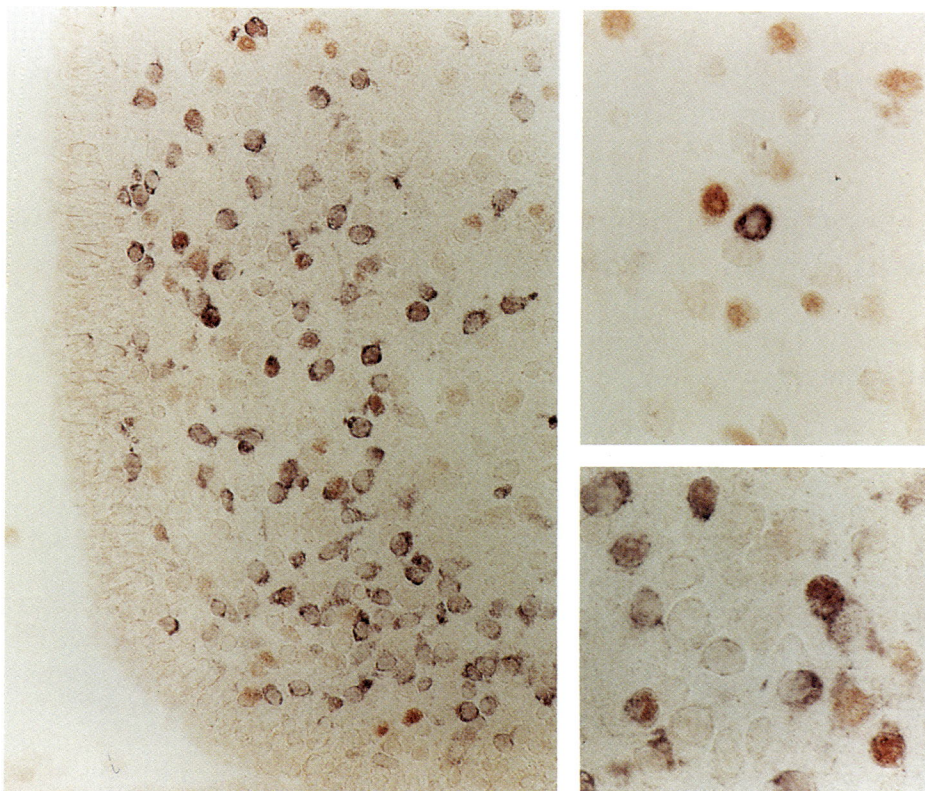

Fig. 1. Left: The arcuate nucleus bordered by the third ventricle. The purple staining is DIG-labeled NPY mRNA, and the brown staining is DAB-labeled Fos. Top
right: Higher magnification of the arcuate nucleus showing individually labeled cells.
Bottom right: Higher magnification of the arcuate nucleus showing examples of single-
and double-labeled cells

problems by using digoxigenin (DIG)-labeled riboprobes for NPY that clearly
label NPY cell bodies without the need for such harsh pretreatments (**Fig. 1**). This
chapter describes a modification and hybrid of protocols (*in situ* hybridization
and immunocytochemistry) developed by Lowry et al. *(1)* and Rogers et al. *(2)*.

## 2. Materials
### 2.1. *DIG* In Situ

*See* **Note 1**.

1. Diethyl pyrocarbonate (DEPC) $H_2O$: 1 mL of DEPC to 1 L of $H_2O$. Stir in hood
   overnight. Autoclave.

2. Phosphate buffer (PB) 0.1 $M$: add 2.7 g sodium phosphate, monobasic and anhydrous, 21.6 g sodium phosphate, dibasic and anhydrous, and 900 mL DEPC $H_2O$ together. Then adjust pH to 7.4 and q.s. to 1 L.
3. Paraformaldehyde 4%: heat 500 mL of PB to 50°C. Add 20 g of paraformaldehyde. Stir until mostly dissolved, approx 30 min. Do not use if temperature exceeded 60°C while making.
4. Triton X-100.
5. Triethanolamine (TEA) 0.1 $M$: 14.9 g TEA, pH to 8.0, and q.s. to 1 L. If TEA is a liquid: 13.27 mL, pH to 8.0, and q.s. to 1 L.
6. TEA 0.1 $M$ with acetic anhydride: 625 µL acetic anhydride in 250 mL 0.1 $M$ TEA.
7. Ethanol 70%.
8. Ethanol 80%.
9. Ethanol 95%.
10. Hybridization solution (100 µL/slide is recommended). To make 4 mL:
    a. 2.5 mL deionized formamide.
    b. 1.0 mL 50% dextran sulfate: 20 g dextran sulfate and q.s. to 40 mL with DEPC $H_2O$.
    c. 300 µL 5 $M$ NaCl: dissolve 292.2 g NaCl in 800 mL of $H_2O$, and then bring volume up to 1 L with $H_2O$.
    d. 40 µL 1 $M$ Tris-HCl, pH 8.0: dissolve 121.0 g Tris base in 1 L $H_2O$; check to make sure the solution is equilibrated to room temperature, then pH to 8.0.
    e. 8 µL 0.5 $M$ EDTA: dissolve 186.1 g EDTA $2H_2O$ in 800 mL $H_2O$ and pH to 8.0 (*see* **Note 2**); and q.s. to 1 L.
    f. 100 µL 50X Denhardt's solution: add 1 g Ficoll, 1 g polyvinylpyrrolidone, and 1 g bovine serum albumin to 1 L of $H_2O$. Filter through a 0.2-µm disposable filter, and store in 10-mL aliquots at –20°C.
    g. 12 mL DEPC $H_2O$.
    Excess hybridization solution can be stored at –20°C and reused.
11. Dithiothreitol (DTT) 1 $M$: dissolve 1.54 g of DTT and 33.5 µL of NaOAc (pH 5.2) in 10 mL $H_2O$. Freeze in 1-mL aliquots. Do not refreeze.
12. tRNA:phenol:chloroform extract and store as 10 mg/mL stock.
13. 20X standard saline citrate (SSC): 175.3 g NaCl, 88.2 g sodium citrate, q.s. to 1 L with $H_2O$, and pH to 7.0.
14. RNase Buffer: 7 mL Tris-HCl (pH 8.0), 70 mL 5 $M$ NaCl, 1.4 mL 0.5 $M$ EDTA, q.s. to 700 mL.
15. RNase A: store as 4 mg/mL stock.
16. 50% EtOH + 300 m$M$ ammonium acetate: 21 mL 5 $M$ ammonium acetate, 185 mL 95% EtOH, q.s. to 350 mL with $H_2O$.
17. 85% EtOH + 300 m$M$ ammonium acetate: 21 mL 5 $M$ ammonium acetate, 313 mL 95% EtOH, q.s. to 350 mL with $H_2O$.
18. Blocking buffer: 2X SSC with 0.05% Tween-20 and 2% normal serum (serum type depends on the animal the anti-DIG is raised in).
19. Buffer 1: 100 mL 1 $M$ Tris-HCl (pH 7.5), 30 mL 5 $M$ NaCl, q.s. to 1 L with $H_2O$.
20. Antibody dilution buffer: buffer 1 with 0.3% Tween-20 and 1% normal serum.

11. Buffer 2: add equal parts of 300 m*M* Tris-HCl, pH 9.5, 300 m*M* NaCl, and 150 m*M* MgCl$_2$ immediately before use.
    a. 300 m*M* Tris-HCl, pH 9.5: 18.17 g Tris-HCl, pH to 9.5, and q.s. to 500 mL with H$_2$O.
    b. 300 m*M* NaCl: 8.88 g NaCl, q.s. with H$_2$O to 500 mL.
    c. 150 m*M* MgCl$_2$ :15.25 g MgCl$_2$, q.s. to 500 mL with H$_2$O.
22. Chromagenic solution: 300 μL of Nitro blue tetrazolium chloride/5-bromo-4-chloro-3-indolyl-phosphate-4-tolnidine salt (NBT/BCIP; Boehringer Mannheim DIG Nucleic Acid Detection Kit #1175041), 75 μL levamisole stock (48 mg/mL), q.s. to 15 mL with buffer 2.
23. 1X Tris-EDTA (TE): 5 mL 1 *M* Tris-HCl, pH 8.0, 1 mL 0.5 *M* EDTA, q.s. to 500 mL with H$_2$O.

## 2.2. Immunocytochemistry

1. 0.1 *M* PBS: add 9 g of NaCl to 1 L of 0.1 *M* PB.
2. 30% H$_2$O$_2$ (make up in 0.1 *M* PBS).
3. 1% H$_2$O$_2$ (make up in 0.1 *M* PBS).
4. PBS+: 0.1 *M* PBS, 0.2% Triton X-100, and 0.1% normal serum (depends on the animal in which the secondary antibody is raised). Make fresh every time, and do not refreeze the serum.
5. Anti-fos antibody.
6. Biotinylated secondary antibody.
7. ABC-Elite standard kit (Vector #PK6100).
8. Diaminobenzidine tetrahydrochloride (DAB) tablets (Sigma #D5905).

## 2.3. NPY Probe

The plasmid containing the NPY sequence was a generous gift from Dr. Denis Baskin. The NPY anti-sense sequence that is contained in the plasmid is as follows:

atatacacgacaacaagggaaatgggtcggaatccagcctggtggtggcatgcattggtgggacaggcag
actggtttcacaggatgagatgagatgtggggggaaactaggaaaagtcaggagagcaagtttcatttcccatc
accacatggaagggtcttcaagccttgttctgggggcattttctgtgctttctctcattaagagatctgaaatca
gtgtctcagggctggatctcttgccatatctctgtctggtgatgagattgatgtagtgtcgcagagcggagtag
tatctggccatgtcctctgctggcgcgtcctacgcccggattgtccggcttggaggggtacccctcagccagaatg
cccaaacacacgagcagggatagagcgagggtcagtccacacagccccattcgtttgttacctagcatcat
ggcgggcgggctgtggtctctgcgcagcgg

# 3. Methods
## 3.1. DIG Labeling

The Boehringer Mannheim DIG RNA labeling kit (#1175025) was used with the following protocol.

1. Combine the following in an RNase/DNase-free 1.5-mL Eppendorf tube:

| DNA | 1 µg linearized for antisense transcription (make sure the enzyme used to digest has been either heat inactivated or passed through an enzyme removal spin column |
|---|---|
| NTP mix | 2 µL |
| 10X transcription buffer | 2 µL |
| RNasin (20 U/mL) | 1 µL |
| T3/T7 or SP6 (20 U/mL) | 2 µL |
| DEPC H$_2$O | up to 21 µL |

2. Incubate for 2 h at 37°C.
3. Add 2 mL of DNase (10 U/µL) and then incubate for 15 min more at 37°C.
4. Stop the reaction, and precipitate or use a Sephadex column (Boehringer Mannheim mini Quick Spin RNA Columns #1 814 427) to remove the excess nucleotides.
5. Add 2.5 µL 4 *M* DEPC LiCl.
6. Add 75 µL 95–100% DEPC EtOH (can be placed on dry ice or –80°C overnight).
7. Spin at 18,300*g* at 4°C 15–30 min.
8. Rinse with 70% DEPC EtOH.
9. Spin at 18,300*g* again for 5 min.
10. Dry in lyophilizer, and resuspend in: 35 µL of DEPC water + 1 µL of RNasin.
11. Sample is ready for concentration determination. To determine the concentration of the probe, perform a dot-blot analysis on the probe using DIG-labeled RNA of a known concentration for a standard.
12. Aliquot 2.5 µL for concentration determination, aliquot the rest in 5-µL aliquots, and store at –80°C until ready to use.

### 3.2 In situ *Hybridization*

*Day 1*

1. Remove slides from –80°C, and allow to air-dry 20 min.
2. Fix frozen tissue in 4% paraformaldehyde for 25 min.
3. Rinse in 0.1 *M* PB and 0.1% Triton X-100 for 10 min.
4. Rinse in 0.1 *M* PB for 5 min.
5. Dip the slides in H$_2$O.
6. Dip in 0.1 *M* TEA.
7. Rinse in 0.1 *M* TEA with acetic anhydride for 10 min. Stir in the acetic anhydride immediately prior to rinsing the slides.
8. Rinse slides in 2X SSC for 3 min.
9. Dehydrate slides in 70%, 80%, and 95% DEPC EtOH for 3 min each.
10. Air-dry slides completely.
11. Add 1 *M* DTT to appropriate amount of hybridization solution (40 µL of 1 *M* DTT/4 mL of hybridization solution). Do not refreeze hybridization solution that has 1 *M* DTT added to it.
12. Combine the DIG-labeled probe (0.3 µg/mL of total hybridization solution) and the tRNA (50 mL of tRNA/mL of hybridization solution) to 50 µL hybridization solution, and heat for 5 min at 85°C to denature the probe.

13. Add the denatured probe and tRNA to the remainder of the hybridization solution, and mix well.
14. Apply 100 μL of probe (in hybridization solution) to each slide, and cover with a Parafilm coverslip.
15. Place all slides into Tupperware containers that have DEPC-moistened paper towels in the bottom (a hybridization chamber), and hybridize at 55°C overnight (*see* **Note 3**).

*Day 2*

1. Rinse in 4X SSC at room temperature twice for 15 min each.
2. Place in RNase buffer + RNase A (30 μg/mL) that has been prewarmed to 37°C for 30 min.
3. Rinse in prewarmed RNase buffer for 30 min at 37°C.
4. Wash in 2X SSC at room temperature for 30 min.
5. Wash in 0.1X SSC that has been prewarmed to 65°C for 30 min.

## 3.3. ICC

*See* **Note 4**.
*Day 1 ICC/ c-Fos*

1. Rinse with 0.1 *M* PBS.
2. Wash in 1% $H_2O_2$ for 10 min.
3. Rinse with PBS four times for 5 min each. If bubbles still remain from the peroxide step, continue with more washes.
4. Apply 100 μL of PBS+ to each slide and cover with a Parafilm coverslip. Set the slides on a level surface, and let them prehybridize for 1 h (*see* **Note 5**). Do not allow the tissue sections to dry out.
5. Remove coverslip, and add 100 μL of PBS+ with antibody. (We use anti-fos at a dilution of 1:500 [*see* **Note 6**], but the dilution factor should be determined for each individual lot number.) Cover slides with a Parafilm coverslip, and place flat in a Tupperware container that has paper towels previously moistened with $H_2O$ across the bottom. Again, the tissue should not be allowed to dry.
6. Let incubate overnight.

*Day 2 ICC/ c-Fos*

1. Rinse with 0.1 *M* PBS 3 times for 5 min each.
2. Add 100 μL of biotinylated donkey anti-rabbit IgG diluted to 1:400 in PBS+ to each slide and cover with a parafilm coverslip for 1 h. Keep tissue moist for the entire hour.
3. Rinse with 0.1 *M* PBS three times for 5 min each.
4. Incubate in a slide mailer of ABC Elite diluted to 1:1500 in PBS for 1 h at room temperature. Make up ABC 30 min before use.
5. Rinse in 0.1 *M* PBS three times for 5 min each.
6. Incubate with DAB (*see* **Note 7**) for 10 min (or until detection is complete) at room temperature.

     10 mg DAB tablet (Sigma)
     50 mL PB
     20 mL $H_2O_2$ (added at last minute)

7. Rinse three times with 0.1 $M$ PB three times for 5 min each time to stop the reaction.

### DIG detection

1. Wash in 0.1X SSC for 3 min.
2. Block in blocking buffer for 60 min.
3. Wash in buffer 1 twice for 10 min each.
4. Apply 100 μL anti-DIG diluted to 1:1000 in antibody dilution buffer to each slide, and cover with a Parafilm coverslip. Keep in a Tupperware container with paper towels that have been moistened with water at 4°C overnight.
5. Remove coverslip, and wash in buffer 1 twice for 10 min each.
6. Wash slides in buffer 2 in slide mailers for 10 min.
7. Detect DIG in chromogen solution for 1–24 h in the dark.
8. Stop detection by washing slides in TE twice for 15 min each.

### Coverslipping

Boehringer Mannheim recommends using Aquapolymount (Polysciences #18606) to coverslip DIG-labeled slides.

## 4. Notes

1. Day 1 *in situ* hybridization reagents must be DEPC treated or made using molecular grade chemicals and DEPC $H_2O$. All the glassware used in day 1 *in situ* hybridization must be baked at 180°C for 6 h or treated with RNaseZap (Ambion #9780) before using.
2. The EDTA will not fully dissolve into solution until the pH is 8.0.
3. There seems to be an interaction between the NBT/BCIP and the DAB, so the DAB detection will only work if it is detected before the DIG.
4. Do not let the tissue sections dry at this stage. If this is a problem, use rubber cement to form a seal between the Parafilm and the slide.
5. The prehybridization will reduce background, but if background is not a problem, this step can be eliminated.
6. The dilutions given for all the antibodies in this protocol are based on specific lot numbers of antibodies. Adjustments may be necessary.
7. DAB should be kept out of the light. Keep covered in foil while dissolving, and wrap the slide mailer in foil during detection.

## Acknowledgments

The authors thank Drs. Michael Lehman, Lique Coolen, Heiko Jansen, and KatiaTsonis for a tremendous amount of sage advice as we developed these methods in our laboratory. We also like thank Paula Stevens for her expert technical advice. This work was supported by two grants to Randy J. Seeley from NIH (DK-54080 and DK-54890).

## References

1. Lowry, C. A., Richardson, C F., Zoeller T. R., Miller L. J., Muske L. E., and Moore F. L. (1997) Neuroanatomical distribution of vasotocin in a urodele amphibian (*Taricha granulosa*) revealed by Immunohistochemical and *In situ* hybridization Techniques. *J. Comp. Neurol.* **385,** 43–70.
2. Rogers, K. V., Vician, L., Steiner, R. A., and Clifton, D. K. (1988) The effect of hypophysectomy and growth hormone administration on preprosomatostatin messenger ribonucleic acid in the periventricular nucleus of the rat hypothalamus. *Endocrinology* **122,** 586–591.

# V

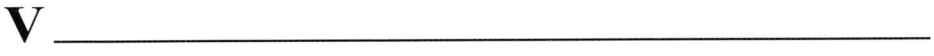

## MISCELLANEOUS TECHNIQUES

# 16

## Y Receptors Characterized by RT-PCR

*Distribution in Rat Intestine*

### Mathieu Goumain, Thierry Voisin, and Marc Laburthe

### 1. Introduction

Receptors for peptide YY (PYY) were discovered in the rat small intestine epithelium *(1)* and were defined as *PYY-preferring* because they display a slightly higher affinity for PYY than for neuropeptide Y (NPY). In contrast, they have a very low affinity for pancreatic polypeptide (PP). They are exclusively present in proliferative crypt cells and appear to be quenched when epithelial cells migrate onto the villi, where they stop to divide and differentiate into mature enterocytes *(2)*. They are not expressed by colonic epithelium *(1)*. Altogether, these observations support the idea that cryptic PYY-preferring receptors may be responsible for the antisecretory actions of PYY and NPY in the small intestine *(3)*.

However, an increasing number of observations (as noted here) contradict this simplistic view: (1) PYY or NPY inhibitory effects on small intestinal secretion are partially blocked by tetrodotoxin *(4)*, supporting the possibility of a neurally mediated mechanism; (2) PYY or NPY inhibit fluid secretion in colon *(5)* despite the absence of expression of PYY receptors in colon epithelium *(1)*; (3) NPY and PYY stimulate intestinal absorption *(6)*, an event that probably occurs in intestinal villi that do not express PYY receptors; and (4) PP as well as certain analogs of PYY and NPY that have a very low affinity for cryptic PYY receptors are nonetheless potent inhibitors of intestinal secretion *(7)*.

Several receptor subtypes belonging to the Y family of heptahelical G-protein-coupled receptors that bind NPY, PYY, and/or PP have been cloned *(8–11)*. Of these, Y1, Y2, and Y5 receptors have similar affinity for NPY and PYY; the Y4 receptor has a high affinity for PP, and the Y6 receptor gene is not expressed in rat *(12)*.

From: *Methods in Molecular Biology, vol. 153: Neuropeptide Y Protocols*
Edited by: A. Balasubramaniam © Humana Press Inc., Totowa, NJ

In this context, we have determined the expression of Y1, Y2, Y4, and Y5 receptors in rat small intestine and colon. For this purpose we have designed highly specific reverse transcriptase-polymerase chain reaction (RT-PCR) primers to monitor the expression of each receptor mRNA in the epithelial (isolated small intestinal crypt cells or villus cells and isolated colonic cells) and nonepithelial layers of gut. Our data provide evidence for a wide distribution of Y receptors in the intestine with differences in their cellular or tissue expression.

## 2. Materials

### 2.1. Isolation and Preparation of Rat Intestinal Tissues

1. Three-month-old male Wistar rats (280–300 g) (IFFA CREDO, L'Arbresle, France) fed ad libitum.
2. NaCl Krebs Ringer's phosphate buffer: 0.118 $M$ NaCl, 5 m$M$ KCl, 1.2 m$M$ MgSO$_4$, 1.2 m$M$ KH$_2$PO$_4$, 10 m$M$ Na$_2$HPO$_4$, pH 7.5, stored at 4°C.
3. Dispersing solution : 0.24 $M$ NaCl and 2.5 m$M$ EDTA, stored at 4°C.
4. A glass rod (diameter approx 3 mm, length approx 50 cm).

### 2.2. RNA Extraction

1. Total RNA extraction kit RNAXEL (Eurobio, Les Ulis, France), stored at 4°C.
2. Chloroform/isoamyl alcohol 24:1 (Interchim, Montluçon, France).
3. Phenol/chloroform/isoamyl alcohol 25:24:1 (Interchim).
4. Isopropanol (ACS grade).
5, Nuclease-free water.
6. 75% ethanol (ACS grade).
7. Eurofine 0.6-mm-diameter needle (Amilabo, Chassieu, France).
8. RQ1 RNase-free DNase (Promega, Charbonnières, France), stored at –20°C.
9. RQ1 DNase 1X buffer: 40 m$M$ Tris-HCl, pH 7.9, 10 m$M$ NaCl, 6 m$M$ MgCl$_2$, 10 m$M$ CaCl$_2$, stored at –20°C.
10. GeneQuant spectrophotometer (Pharmacia Biotech, Orsay, France).

### 2.3. RT-PCR Experiments

1. M-MLV reverse transcriptase (Life Technologies, Cergy Pontoise, France), stored at –20°C.
2. M-MLV reverse transcriptase 5X buffer (Life Technologies), stored at –20°C : 250 m$M$ Tris-HCl, pH 8.3, 375 m$M$ KCl, 15 m$M$ MgCl$_2$.
3. Dithiothreitol (DTT) 0.1 $M$ (Life Technologies), stored at –20°C.
4. Oligo (dT)$_{15}$ (Promega), stored at –20°C.
5. dNTPs 10 m$M$ each (Promega), stored at –20°C.
6. Taq DNA polymerase 5 U/μL (Life Technologies), stored at –20°C.
7. Taq DNA polymerase 10X PCR Buffer (Life Technologies), stored at –20°C: 200 m$M$ Tris-HCl, pH 8.4, 500 m$M$ KCl.
8. PCR primers (chosen with OligoStructure 3.3 PCR primer search program), stored at –20°C:

| | |
|---|---|
| Y1 subtype (rat/mouse) | PCR product: 546 bp |
| Sense (pos. 499) | 5' GCT TCT TCT CTG CCC TTY GTG 3' |
| Antisense (pos. 1044) | 5' RGT CTC GTA GTC RTC GTC TCG 3' |
| Y2 subtype (human) | PCR product: 442 bp |
| Sense (pos. 349) | 5' AAA TGG GTC CTG TCC TGT GCC 3' |
| Antisense (pos. 790) | 5' TGC CTT CGC TGA TGG TAG TGG 3' |
| Y4 subtype (rat/mouse) | PCR product: 492 bp |
| Sense (pos. 623) | 5' GAC TTG CTA CCC ATC CTC ATM 3' |
| Antisense (pos. 1115) | 5' ATC ACC ACC GYC TCA TCT AYA 3' |
| Y5 subtype (rat) | PCR product: 524 bp |
| Sense (pos. 834) | 5' CCA GGC AAA AAC CCC CAG CAC 3' |
| Antisense (pos. 1357) | 5' GGC AGT GGA TAA GGG CTC TCA 3' |
| GAPDH (human) | PCR product: 983 bp |
| Sense (pos. 11) | 5' TGA AGG TCG GAG TCA ACG GAT TTG GT 3' |
| Antisense (pos. 992) | 5' CAT GTG GGC CAT GAG GTC ACA CAC 3' |

9. 50 mM $MgCl_2$ (Life Technologies, Cergy Pontoise, France), stored at –20°C.
10. Ethidium bromide 0.625 mg/mL (Interchim, Montluçon, France), stored at room temperature.
11. 0.5X TBE (Interchim, Montluçon, France), stored at room temperature.
12. Denaturing buffer A: 0.5 $M$ NaOH, 1.5 $M$ NaCl, stored at –20°C.
13. Fixing buffer B: 0.5 $M$ Tris-HCl, 1.5 $M$ NaCl, stored at –20°C.
14. Washing buffer C: 3X sodium chloride sodium phosphate EDTA buffer (SSPE), 2 m$M$ EDTA, stored at –20°C.
15. Hybond-N+ positively charged nylon membrane 2.0 (Amersham).
16. Trans-blot SD semi-dry transfer cell (Bio-Rad, Ivry sur Seine, France).
17. Transfer buffer : 0.5X TBE.
18. Rediprime DNA labeling system (Amersham).
19. Hybridization solution: 5X SSPE, 5X Denhardt's solution, 0.5% sodium dodecyl sulfate (SDS), 500 µL denaturated salmon-sperm DNA 0.25 mg/mL, stored at –20°C.
20. Kodak Biomax MR film (Eastman Kodak, Rochester, NY).

## *2.4. PCR Product Cloning and Sequencing*

1. The LigATor Kit (R&D systems, Abingdon, UK) containing 10X ligase buffer (200 m$M$ Tris-HCl, pH 7.6, 50 m$M$ $MgCl_2$), 10 m$M$ ATP, T4 DNA ligase (100 U), 100 m$M$ DTT, SOC medium, competent cells, stored at –20°C.
2. 100 m$M$ isopropyl β-ᴅ-thiogalactopyranoside (IPTG) in $H_2O$, stored at –20°C.
3. 20 mg/mL 5-bromo-e-chloro-3-indolyl-β-ᴅ-galactoside (X-gal) in dimethyl sulfoxide, stored at –20°C.
4. 50 mg/mL ampicillin in $H_2O$, stored at –20°C.
5. Nucleobond AX 100 Kit (Macherey-Nagel, Hoerdt, France).
6. Sequenase Version 2.0 DNA Sequencing Kit (Amersham).

## 3. Methods

### *3.1. Isolation and Preparation of Rat Intestinal Tissues*

1. Remove rat jejunum and colon after decapitation at 9–10 AM. Also collect hypo-thalamus, lungs, and testis.
2. Flush intestinal segments free of contents, place on a glass rod (3 mm diameter), and evert.
3. Fill intestinal segments with NaCl Krebs Ringer's phosphate buffer (4°C).
4. Release villus cells by a brief shaking of the jejunal segment in the dispersing solution (4°C). Following this, prolonged shaking of the jejunal segment in the dispersing solution (4°C) releases crypt cells.
5. Collect epithelial cells from the colon following the same method as for the jejunum.
6. Tissues remaining after complete removal of epithelial cells from jejunal and colon are essentially nonepithelial tissues.
7. Store cells and tissue samples at 4°C until total RNA extraction.

### *3.2. RNA Extraction*

1. Recover cells in suspension by centrifugation at 800*g* for 10 min. Resuspend the cell pellets in the RNAXEL extraction buffer to give approx 1 mL for every 100 mg tissue. Shear the cells by passage through a needle (diameter 0.6 mm) 10 times. For solid tissues, including nonepithelial intestine, nonepithelial colon, hypothalamus, lung, or testis, homogenize in 1 mL RNAXEL/100 mg of tissue using an Ultra-Turrax.
2. Extract the RNA using chloroform (0.2 mL for a 1-mL suspension), and vortex vigorously. Leave on ice for 5 min and centrifuge at 11,000*g* for 15 min at 4°C.
3. Carefully recover the aqueous phase (top phase) in a sterile plastic tube, and add 0.5 vol isopropanol and 0.05 vol of RNABIND resin. Vortex for 30 s, and centrifuge at 11,000*g* for 1 min at 4°C.
4. Discard the supernanant, and wash the pellet twice in 1 mL 75% ethanol by 30-s vortexing followed by 30-s centrifugation. Discard the supernatant and dry the pellet to remove ethanol completely.
5. Add 0.1 vol of nuclease-free water. Vortex for 30 s, centrifuge for 1 min, and transfer the supernatant to a clean tube. Repeat this step.
6. Quantify the yield of RNA using a spectrophotometer by measuring the absorbance at 260 nm and 280 nm. The ratio of absorbance at 260 and 280 nm should exceed 1.8.
7. To prevent DNA contamination during the RNA extraction process, treat the samples with RQ1 RNase-free DNase as follows (*see* **Note 1**): to 15 μL of total RNA in 75 μL nuclease-free water, add 10 μL RQ1 DNase 10X buffer and 15 μL RQ1 DNase (15 U). Incubate for 1 h at 37°C.
8. Perform phenol/chloroform (v/v) extraction and 75% ethanol precipitation. Resolubilize RNA in ultrapure water at a concentration of 1 μg/μL (*see* **Note 2**).

## 3.3. RT-PCR Experiments

1. Dilute 10 µg of total RNA in 12 µL nuclease-free water, and denature by heating for 5 min at 75°C. Add: 2 µL 0.1 $M$ DTT, 4 µL M-MLV 5X buffer, 1 µL dNTP, 0.5 µL oligo(dT)$_{15}$ (100 µ$M$), and 0.5 µL M-MLV (100 U).
2. Incubate for 1 h at 37°C and heat 5 min at 72°C.
3. Make up a bulk PCR mix (45 µL x number of cDNA samples) containing 0.5 µL of each chosen primer, 0.5 µL dNTP, 1.5 µL 50 m$M$ MgCl$_2$, 5 µL Taq DNA polymerase 10X buffer, 0.5 µL Taq DNA polymerase (2.5 U) and H$_2$O up to 45 µL. Add 5 µL of the cDNA mixture (obtained in **Subheading 3.3.2.**) to 45 µL of PCR mix.
4. Perform PCR for 35 cycles as follows: denaturated at 92°C for 1 min; annealed at 64°C for 1 min ; extended at 72°C for 1 min.
5. Following PCR, add 2.5 µL of 5X gel loading dye to each tube, and electrophorese the samples on agarose gel (1%) with ethidium bromide (1 drop for 50 mL) in 0.5X TBE.
6. Immerse the agarose gel for 30 min in denaturing buffer A, 15 min in fixing buffer B, and 15 min in washing buffer C.
7. Set up transfer assembly in the following order: three pieces of Whatman 3MM paper, nylon N+ membrane, the gel, and finally a further three pieces of 3MM paper.
8. Perform the transfer of DNA by semi-dry electro-blotter for 15 min at 3.55 mA/cm$^2$ of gel area, using 0.5X TBE as transfer buffer.
9. Radiolabel the Y receptor cDNA using the Rediprime DNA labeling system (Amersham, Life Science) as follows:
   a. Dilute the DNA to be labeled to a concentration of 2.5–25 ng in 45 µL of sterile water.
   b. Denature the DNA sample by heating to 100°C for 5 min in a boiling water bath.
   c. Centrifuge briefly to bring the contents to the bottom of the tube.
   d. Add the denatured DNA to the labeling mix, and reconstitute the mix by gently flicking the tube until the blue color is evenly distributed.
   e. Centrifuge briefly, add 5 µL of [α$^{32}$P]dCTP (5.10$^6$ cpm/mL), and mix by gently pipeting up and down four to five times.
   f. Centrifuge briefly, and incubate at 37°C for 10 min.
10. Prehybridize the N+ nylon filter for 4 h at 65°C in hybridization solution.
11. Add the radiolabeled probe (after a 5-min denaturation step at 100°C), and hybridize overnight at 65°C.
12. Briefly rinse the filter in 2X SCC, 0.1% SDS at room temperature.
13. Successively wash the filter twice for 10 min with 20 mL of 2X SCC, 0.1% SDS; 1X SCC, 0.1% SDS; 0.1X SCC, 0.1% SDS.
14. Dry the filter, and fix the DNA to the filter with a 5-min UV exposure.
15. Place the filter in contact with the emulsion side of a Kodak film and expose for 1–24 h.

### 3.4. PCR Product Cloning and Sequencing

1. Set up ligation reaction: 2 µL PCR fragment (0.2 pmol); 1 µL 10X ligase buffer; 0.5 µL 100 m*M* DTT; 0.5 µL 10 m*M* ATP; 1 µL pTAg vector.
2. Add $H_2O$ to make up to 9.5 µL microcentrifuge, vortex, microcentrifuge, add 0.5 µL T4 DNA ligase (2–3 Weiss units), and mix gently.
3. Incubate overnight at 16°C.
4. Thaw LigATor's competent cells gently at 4°C. Mix carefully and pipet 20 µL into a precooled tube. Thaw SOC medium at room temperature.
5. Add 1 µL ligation reaction to cells. Tap tube, and leave on ice for 30 min.
6. Heat shock at 42°C for exactly 40 s, and incubate on ice for 2 min.
7. Add 80 µL SOC medium and shake at 200–250 rpm in a rotary shaking incubator for 1 h at 37°C.
8. Plate 50 µL onto Luria-Bertani (LB) agar plates containing 40 µL X-gal, 100 µL IPTG, and 0.05 mg/mL ampicillin.
9. Incubate plates inverted at 37°C overnight in a humidified incubator. More intense blue color development may be observed if the plates are placed at 4°C for 2–3 h following incubation at 37°C.
10. Select 5–10 white colonies, and transfer them using a pipet end to 30 mL LB with 0.05 mg/mL ampicillin. Grow at 37°C for 12–15 h.
11. DNA mini-preparation (using Nucleobond AX100 cartridge; Macherey-Nagel):
    a. Suspend bacterial cell pellet carefully in 4 mL of buffer S1.
    b. Add 4 mL of buffer S2, and mix the suspension gently by inverting tube.
    c. Incubate the mixture at room temperature for 5 min.
    d. Add 4 mL of buffer S3, and mix gently until a homogenous suspension is formed; then incubate on ice for 10 min.
    e. Equilibrate the cartridge with 2 mL of buffer N2.
    f. Centrifuge the suspension for 30 min at 20,000*g* at 4°C, and then load the supernatant onto the preequilibrated cartridge.
    g. Wash the cartridge with 4 mL of buffer N3. Repeat washing one more time.
    h. Elute the plasmid DNA with 2 mL of buffer N5. Repeat elution step one more time.
    i. Precipitate plasmid DNA with 0.7 vol of isopropanol, preequilibrated to room temperature. Centrifuge at high speed (>11,000*g*) at 4°C. Wash DNA with 70% ethanol, dry briefly, and redissolve in water at 1 µg/mL.
12. Sequence plasmid inserts by a dideoxy chain-termination method, using the commercially available kit SEQUENASE 2.0 (Amersham, Life Science).

## Discussion

In the present study, rat intestinal expression of the recently cloned Y1, Y2, and Y5 receptors (with a similar affinity for PYY and NPY) and Y4 receptors (with a high affinity for PP) was investigated by RT-PCR *(13)*. The data show that all Y receptors are expressed in small intestine and/or

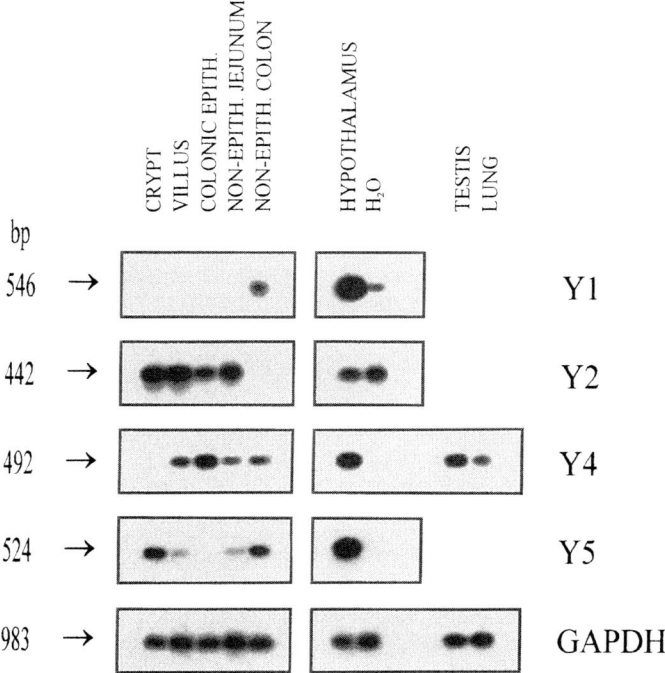

Fig. 1. PCR-based detection of mRNA expression of Y1, Y2, Y4, and Y5 receptors in intestine, colon, and control tissues. cDNAs from indicated tissues were used as templates for PCR reactions using specific primers (*see* text). Southern hybridization with specific [$^{32}$P]-labeled probes is shown. Colonic epith., colonic epithelium; Non-epith. jejunum, nonepithelial jejunum ; Non-epith. colon, nonepithelial colon.

colon but with specific distributions (**Fig. 1**). Y1 receptors are only expressed in nonepithelial colonic tissue, whereas Y2 and Y4 receptors are present in both epithelial and nonepithelial tissues of the small or large intestine. In contrast, Y5 receptor expression appears to be restricted to epithelial crypts of the small intestine and nonepithelial tissue of the colon. Sequencing of PCR products showed 100% identity with the corresponding sequences of the cloned Y1, Y4, or Y5 receptors. The PCR product obtained with Y2 primers from rat crypt cells showed 84% identity with the cloned human Y2 receptor (*14*). These data indicate a wide distribution of Y receptors in the small intestine and colon. They also suggest that Y1, Y2, Y4, and Y5 receptors may be responsible for still unexplained effects of PYY, NPY, and/or PP on secretion or absorption in the small and large intestines.

## 4. Notes

1. Total RNA can be stored in 75% ethanol at –80°C until RT-PCR experiments. In that case, total RNA has to be precipitated for 1 h to overnight at –80°C with 0.3 $M$ Na acetate buffer, pH 6.0, followed by a 30-min centrifugation (11,000$g$). The total RNA pellet is then washed with 75% ethanol, briefly dried, and redissolved in RNase-free water at 1 µg/µL.
2. DNase treatment ensures that the product resulting from RT-PCR is due to amplification of cDNA that has been synthesized during the reverse transcriptase reaction. This is an important step, particularly to study genes devoid of introns. An alternative procedure can be tried by replacing RQ1 DNase buffer with $MgCl_2$ at a final concentration of 2 m$M$.

## References

1. Laburthe, M., Chenut, B., Rouyer-Fessard, C., Tatemoto, K., Couvineau, A., Servin, A., et al. (1986) Interaction of peptide YY with rat intestinal epithelial plasma membranes: binding of the radioionated peptide. *Endocrinology* **118,** 1910–1917.
2. Voisin, T., Rouyer-Fessard, C., and Laburthe, M. (1990) Distribution of the common peptide YY/neuropeptide Y receptor along rat intestinal villus-crypt axis. *Am. J. Physiol.* **258,** G753–G759.
3. Laburthe, M. (1990) Peptide YY and neuropeptide Y in the gut: availability, biological actions and receptors. *Trends Endocr. Metab.* **1,** 168–174.
4. Souli, A., Chariot, J., Presset, O., Tsocas, A., and Rozé, C. (1997) Neural modulation of the antisecretory effect of the peptide YY in the rat jejunum. *Eur. J. Pharmacol.* **333,** 87–92.
5. Strabel, D. and Diener, M. (1995) The effect of neuropeptide Y on sodium, chloride and potassium transport across the rat distal colon. *Br. J. Pharmacol.* **115,** 1071–1079.
6. MacFadyen, R. J., Allen, J. M., and Bloom, S. R. (1986) NPY stimulates net absorption across rat intestinal mucosa in vivo. *Neuropeptides* **7,** 219–227.
7. Souli, A., Chariot, J., Voisin, T., Presset, O., Tsocas, A., Balasubramaniam, A., et al. (1997) Several receptors mediate the antisecretory effect of peptide YY, neuropeptide Y and pancreatic polypeptide on VIP-induced fluid secretion in the rat jejunum in vivo. *Peptides* **18,** 551–557.
8. Laburthe, M., Couvineau, A., and Voisin, T. (1999) Receptors for peptides of the VIP/PACAP and PYY/NPY families, in *Gastrointestinal Endocrinology* (Greeley, G. H., ed.), Humana, Totowa, NJ, pp. 125–158, R.
9. Blomqvist, A. G. and Herzog, H. (1997) Y-receptor subtypes—how many more? *Trends Neurosci.* **20,** 294–298.
10. Larhammar, D. (1996) Structural diversity of receptors for neuropeptide Y, peptide YY and pancreatic polypeptide. *Regul. Pept.* **65,** 165–174.
11. Cox, H. M. (1998) Peptidergic regulation of intestinal ion transport. A major role for neuropeptide Y and the pancreatic polypeptides. *Digestion* **59,** 395–399.
12. Burkhoff, A., Linemeyer, D. L., and Salon J. A. (1998) Distribution of a novel

hypothalamic neuropeptide Y receptor gene and it's absence in rat. *Mol. Brain Res.* **53**, 311–316.

13. Goumain, M., Voisin, T., Lorinet, A. M., and Laburthe, M. (1998) Identification and distribution of mRNA encoding the Y1, Y2, Y4 and Y5 receptors for peptides of the PP-fold family in the rat intestine and colon. *Biochem. Biophys. Res. Commu.* **247**, 52–56.

14. Gerald, C., Walker, M. W., Vaysse, P. J., He, C., Branchek, T. A., and Weinshank, R. L. (1995) Expression cloning and pharmacological characterization of a human hippocampal neuropeptide Y/peptide YY Y2 receptor subtype. *J. Biol. Chem.* **270**, 26,758–26,761.

# 17

## Quantification of NPY mRNA
## by Ribonuclease Protection Assay

**Abhiram Sahu**

### 1. Introduction:

The ribonuclease protection assay (RPA) is one of the widely used techniques for detection and quantification of mRNA. This assay offers at least 5–10 times the sensitivity of Northern blots and also enables the investigator to use multiple probes in a single assay. Ribonuclease protection involves the hybridization in a solution of radiolabeled antisense RNA probe to an RNA sample. The antisense RNA will only hybridize to the mRNA of interest, and the unhybridized single-stranded RNA is removed by RNase digestion. The protected duplex radiolabeled hybrid fragments are recovered by ethanol precipitation. The protected fragments are subsequently separated on a denaturing urea-polyacrylamide gel and analyzed by autoradiography. Multiple probes can be used to detect levels of multiple target mRNAs. To normalize the variation between sample loading and losses during various steps, a probe for housekeeping gene, e.g., cyclophilin, glyceraldehyde-3-phosphate dehydrogenase, β-actin, or ribosomal RNA is used. Depending on the experimental condition one has to choose an appropriate housekeeping gene whose mRNA levels are not altered by the experimental manipulation. In this chapter I present the RPA technique for quantification of prepro-neuropeptide Y (NPY) mRNA levels in the hypothalamic tissue. This technique has been adapted in our laboratory from methods described elsewhere *(1,2)*. The procedure can be broken down into five parts: (1) total RNA extraction from the hypothalamic tissue; (2) synthesis of sense RNA and radiolabeled antisense RNA probe; (3) solution hybridization and RNase digestion; (4) separation of protected fragment by gel electrophoresis; and (5) analysis of protected bands by autoradiography or phosphor imaging.

From: *Methods in Molecular Biology, vol. 153: Neuropeptide Y Protocols*
Edited by: A. Balasubramaniam © Humana Press Inc., Totowa, NJ

## 2. Materials

### 2.1. Total RNA Isolation

1. Diethyl pyrocarbonate (DEPC)-$H_2O$: add 10 μL DEPC per 100 mL Milli-Q water, stir for >15 min, and autoclave for 20 min to eliminate residual DEPC.
2. Extraction buffer: 4 $M$ guanidine isothiocynate, pH 7.0, 25 m$M$ Na citrate, 0.5% sodium-$N$-lauryl sarcosin, 0.1 $M$ β-mercaptoethanol. This buffer is light sensitive. Keep wrapped with aluminum foil at room temperature, and use within 1 week.
3. Phenol equilibrated with Tris-EDTA buffer, pH 8.0. Store at 4°C. For long-term storage (more than 6 months), keep at –20°C in a brown plastic bottle, and bring to 4°C on the day before RNA extraction.
4. Chloroform/isoamyl alcohol (24:1).
5. 2 $M$ NaOAc (sodium acetate).
6. 5 $M$ $NH_4OAc$ (ammonium acetate).
7. 95% ethyl alcohol (EtOH).
8. 5X TBE: 0.89 $M$ Tris base, 0.89 $M$ boric acid, 0.02 $M$ EDTA, pH 8.0.
9. Agarose, ethidium bromide (EtBr, light sensitive, carcinogenic).
10. Dye II (10X): 0.25% bromophenol blue, 0.25% xylene cyanol, 25% Ficoll in $H_2O$. Mix, DEPC treat, and autoclave briefly at lower than usual temperature.
11. 1X TE buffer: 10 m$M$ Tris-HCl, 1 m$M$ EDTA, pH 8.0.

### 2.2. RNA Probe Preparation

1. Plasmid with cDNAs for preproNPY mRNA and cyclophilin mRNA. I received plasmids with cDNAs for rat preproNPY mRNA and cyclophilin mRNA from Drs. S. Sabol (NIH) and J. L. Roberts (Mount Sinai School of Medicine, New York) respectively. However, these cDNAs could also be synthesized by reverse transcriptase polymerase chain reaction (RT-PCR).
2. Riboprobe 5X transcription buffer (Stratagene).
3. Dithiothreitol (DTT) 1 $M$, RNasin (RNase inhibitor).
4. NTP: a mixture of 2.5 m$M$ each of triphosphates, ATP, GTP, CTP.
5. $^{32}P$-labeled UTP (Amersham #PB.10163), UTP (2.5 m$M$ ).
6. T3, T7 RNA polymerases (Stratagene).
7. RQ1 DNase, yeast tRNA (as carrier).
8. Phenol, chloroform-isoamyl alcohol (24:1).
9. 5 $M$ $NH_4OAc$, 100% EtOH.
10. Elution buffer: 0.5 $M$ $NH_4OAc$, 1 m$M$ EDTA, 0.1% sodium dodecyl sulfate (SDS), 10 m$M$ Tris-HCl, pH 7.5.

### 2.3. Solution Hybridization and RNase Protection

1. Hybridization buffer: 40 m$M$ PIPES (piperazine-$N$,$N$1-bis-(2-ethane-sulfonic acid), pH 6.5, 400 m$M$ NaCl, 1 m$M$ EDTA, 80% (v/v) deionized formamide (mix 50 mL of formamide and 5 g of mixed-bed ion exchange resin [Bio-Rad #AG 501-X8, 20–50 mesh]. Stir for 30 min at room temperature [RT]. Filter twice

through no. 1 filter paper [Whatman]. Dispense into 1-mL aliquots, and store at –20°C). Make fresh buffer each time.

2. Ribonuclease A/T1 digestion buffer: 10 m$M$ Tris-HCl, pH 7.5, 5 m$M$ EDTA, 300 m$M$ NaCl, with sterile water (do not use DEPC H$_2$O), 40 µg/mL RNase A (Boehringer Mannheim), and 600 U/mL RNase T1 (Life Technologies). Digestion buffer may be kept at RT for 1 month, but add RNase A and T1 only on the day of digestion (*see* **Note 1**).

3. Proteinase K (make fresh), 20% SDS.

4. Chloroform/isoamyl alcohol (24:1), 2-propanol.

5. Yeast tRNA.

## 2.4. Gel Electrophoresis and Autoradiography

1. 80% formamide dye: 3.9 mL deionized formamide, 110 µL 5X TBE (0.89 $M$ Tris, 0.89 $M$ boric acid, 0.02 $M$ EDTA), 5 mg bromophenol blue, 5 mg xylene cyanole FF, 865 µL DEPC H$_2$O.

2. 6% polyacrylamide-8 $M$ urea gel: 24 g urea, 7.5 mL polyacrylamide (38% acrylamide: 2% Bis, keep at 4°C), 10 mL 5X TBE, 50 µL 25% ammonium persulfate (APS, fresh weekly, wrapped with aluminum foil), approx 15 mL Milli-Q water to make 50 mL. Add 50 µL TEMED ($N$,$N$,$N'$,$N'$-tetramethyl-ethyl-enediamine) before pouring gel solution to plates.

3. Gel-soaking solution: 10 mL methanol, 5 mL glycerin, 85 mL H$_2$O.

4. Blotting paper, cellophane paper (Bio-Rad).

5. X-ray film (Kodak X-Omat AR film or Fuji RX film).

# 3. Methods

## 3.1. Total RNA Isolation

### 3.1.1. Extraction of RNA (see **Note 2**)

1. Total RNA from the medial basal hypothalamus (MBH) fragment is isolated by the guanidine isothiocyanate method *(3)*. Fresh or frozen tissues are homogenized in 400 µL of extraction buffer with a Polytron tissue homogenizer (Brinkmann, Piscataway, NJ; or PRO Scientific, Monroe, CT) directly in 1.5-mL microfuge tubes. Homogenize about 15 s, buffer will become frothy—be sure the tissue is completely homogenized.

2. After homogenization, into this tube, add sequentially 40 µL 2 $M$ NaOAc, 400 µL phenol, and 80 µL chloroform/isoamyl alcohol (24:1), with vortexing after each addition. Keep the mixture on ice for 15 min, followed by centrifugation at 12,000$g$ in a microfuge at 4°C.

3. Remove the aqueous (upper) phase slowly, being very careful not to take any white precipitate from the interphase, and precipitate with an equal volume of isopropanol overnight at –20°C.

4. Next day, spin the tubes for 20 min in a microfuge at 4°C, decant carefully, saving the pellet, and dry around pellets with a Kimwipe (*see* **Note 3**).

5. Dissolve the pellets in 50 μL of extraction buffer, and precipitate with 50 μL of isopropanol for 1 h at 4°C.

6. Spin the tubes in a microfuge for 20 min at 4°C, dry around the pellets with a Kimwipe, and rinse twice with 500 μL of 70% alcohol.

7. Dry the pellets in a Speed-Vac (Savant) for 5 min or more as needed, and then dissolve in 35 μL of DEPC H₂O.

8. Determine the RNA concentration by measuring absorbance at 260 nm in a spectro-photometer (Perkin-Elmer). Also calculate the ratio of absorbance at 260 and 280 nm to check the purity of RNA (it should be approx 1.8–2.0).

9. The integrity of the RNA is checked by visualization of the ethidium bromide-stained 28s and 18S ribosomal RNA bands electrophorosed in 1% agarose gel (*see* below).

10. Store RNA samples at –20°C.

### 3.1.2. Agarose Gel Electrophoresis

1. Combine 0.4 g agarose, 8 mL 5X TBE, and 32 mL DEPC H₂O in a conical flask with a stir-bar and aluminum foil over the top. Boil the solution, cool to approx 65°C, add 4 μL of 5 mg/mL EtBr solution in a hood, and swirl the flask to mix.

2. Pour the solution into a mini-gel support, being careful to avoid bubbles, then insert appropriate comb, and allow the gel to solidify. Take extra caution during the use of EtBr; do not inhale any vapor, and always prepare gel in a chemical hood.

3. After the gel is solidified, remove the comb, and place the gel in electrophoresis apparatus.

4. Fill the electrode reservoirs with running buffer (1X TBE) until it covers the gel and touches the electrodes.

5. Load RNA samples (2 μL RNA, 2 μL dye II 10X, 4 μL 1X TE buffer) and *Escherichia coli* rRNA standards, and electrophorose at approx 70 mV at 4°C for 1¹/₂ h, or until the dye-front nears the end of the gel.

6. Turn off the power supply, and photograph the gel using ultraviolet (UV) illumination.

## 3.2. cDNA and RNA Probe Preparation

### 3.2.1. Linearization of the Plasmid

1. Add in a 1.5-mL microfuge tube: 50 μg/35 μL plasmid, 5 μL buffer 3 (Stratagene), 5 μL bovine serum albumin (BSA; Stratagene), and 5 μL restriction enzyme (Stratagene), microfuge for 30 s, vortex and incubate at 37°C overnight.

2. Next day, add 25 μL phenol and 25 μL chloroform/isoamyl alcohol (24:1), vortex, and microfuge for 3 min at RT.

3. Collect the aqueous phase (approx 50 μL) in a new tube, add 50 μL 5 *M* NH₄OAc and 250 μL 100% ethanol (cold), and precipitate for 30 min at –20°C.

4. Microfuge for 10 min at 4°C, wash pellet with 500 μL of 70% alcohol, dry in a Speed-Vac for 3–5 min, and resuspend in 50 μL of DEPC H₂O.

5. Confirm completion of the linearization by electrophoresis in 1% agarose gel (*see* above). Electrophorose 1 μg cut and 1 μg uncut plasmid in the same gel.

*Pvu*II restriction enzyme is used to linearize NPY plasmid to synthesize both antisense cRNA probe and sense RNA standard. To make antisense and sense RNA for cyclophilin, the plasmid is linearized with *Eco*RI and *Bam*HI, respectively.

## 3.2.2. cRNA Probe Preparation

1. $^{32}$P-Labeled NPY cRNA probe is synthesized in vitro using the following protocol. Add in a 0.5-mL microfuge tube: 1 µg/µL linearized plasmid, 1 µL 1 *M* DTT, 1 µL RNasin, 1 µL NTP, 3 µL riboprobe 5X transcription buffer, 8 µL [$^{32}$P]UTP, and 1 µL T3 RNA polymerase. Close cap, quick spin for 30 s, mix gently by flicking (do not vortex), and incubate for 45 min at 37°C.
2. After incubation, digest the DNA template by adding 1 µL RQ1 DNase to the reaction mixture, mix, and incubate at 37°C for 15 min (*see* **Note 4**).
3. To stop this reaction, add 3 µL yeast tRNA, 10 µL phenol, and 10 µL chloroform, vortex, and spin the tube for 5 s in a microfuge at room temperature.
4. Transfer the upper aqueous phase (approx 25 µL) to a new tube, and add 25 µL of chloroform to it, vortex, and spin for 5 s in a microfuge.
5. Remove the aqueous phase, and repeat this step once.
6. Transfer supernatant to a new tube, and precipitate with 20 µL of 5 *M* NH$_4$OAc and 100 µL of 100% EtOH (cold) at −20°C for 30 min.
7. Spin the tube in a microfuge for 5 min, remove the supernatant, and dissolve the pellet in 50 µL DEPC H$_2$O.
8. Precipitate the RNA a second time with 50 µL 5 *M* NH$_4$OAc and 250 µL 100% ETOH (cold) at −20°C for 30 min.
9. After centrifugation for 7 min in a microfuge at 4°C, remove supernatant, and wash the pellet with 300 µL 70% alcohol.
10. Finally, dry around the pellet with a Kimwipe, and dissolve in 100 µL DEPC H$_2$O. Dilute 5 µL of this stock solution to 500 µL, and use 5 µL of this diluted solution with 2 mL of scintillation cocktail (Complete Counting Cocktail 3a70B, Research Products International, Mount Prospect, IL) to count radioactivity in a β-counter.
11. For preparation of cyclophilin cRNA probe, follow the same technique using 5 µL [$^{32}$P]UTP instead of 8 µL, T7 RNA polymerase, and 3 µL RNase free water.
12. After DNA digestion (step 2), the incorporation of [$^{32}$P]UTP into the newly transcribed RNA is determined using DE-81 filter (Whatman) (4). Spot 1 µL of 20X diluted reaction mixture on each of two DE-81 filters. Use one for the total count, and wash the other filter in a vacuum filtering apparatus with 10 mL each of 0.5 *M* sodium phosphate buffer and 95% alcohol.
13. Place the filters in a scintillation vial, add 2 mL scintillation cocktail, and count in a β-counter. Incorporation should be at least 30–40% of the total cpm used.

## 3.2.3. Gel Purification of the Radiolabeled Probe

*See* **Note 5.**

1. For a preproNPY cRNA probe, it is preferable to purify the probe by gel electrophoresis. To do this, prepare a 6% polyacrylamide-8 *M* urea gel (as described in **Subheading 3.4.**).

2. Add 80% formamide dye (gel loading dye) to the probe, heat for 5 min at 95°C, and then place in ice water for 30 s.
3. Load $3 \times 10^6$ cpm equivalent of probe in 5 µL of 80% formamide dye.
4. Electrophorese gel at 50–55°C until bromphenol blue (darker blue) reaches bottom of the gel (approx 2 h).
5. Remove notched plate; cover gel and bottom plate with plastic wrap.
6. Put Scotch Tape at the corners of the gel on the plastic wrap, and mark 2–3 dots with Magic Marker on each piece of tape.
7. Using some old probe, spot 1 µL/dot on the tape, and allow to dry (5–10 min).
8. Place a second piece of tape over the first, keep the gel in a cardboard film holder, and expose to X-ray film for 5–10 min.
9. Using the dots as a guide, line up and invert the X-ray film and the gel, and mark the position of the probe band with a felt pen on the back of the glass.
10. Turn the gel back over, remove the film, and excise out with razor blade or scalpel the area of the gel that contains the full-length transcript (slow migrating intense band on the autoradiograph) indicated by the reference mark on the glass plate.
11. Place the gel piece in a 1.5-mL microfuge tube, add 500 µL of elution buffer, 2 µL RNasin, and mash the gel using a Kontes pellet pestle. Incubate the tube at 37°C for approx 2 h or overnight.
12. Centrifuge tubes for approx 30 s. Recover as much liquid as possible from the tube, and transfer to a fresh tube.
13. Extract with phenol and chloroform (for 400 µL buffer, add 200 µL each of phenol and chloroform/isoamyl alcohol).
14. Transfer the supernatant into a fresh tube, and extract twice with an equal volume of chloroform/isoamyl alcohol (24:1). Count the aqueous phase.
15. Use the probe for hybridization as soon as possible (within the same day; *see* **Note 6**).
16. Reexpose the gel to verify that the film and gel were correctly aligned and that the intended probe band was cut properly.

### 3.2.4. Preparation of Sense RNA for Standards

1. Sense RNA for standards is synthesized essentially in the same manner as the antisense probe, except for changes noted here. For NPY sense RNA transcript, the plasmid linearized with *Pvu*II is transcribed using T7 RNA polymerase in the presence of NTP and cold UTP. All components of the transcription reaction are increased by a factor of 3, and incubation time is increased to 1 h so a measurable amount of sense RNA is synthesized. Also added into the reaction is 2 µL of diluted (1:10) [$^{32}$P]UTP so that the full-length transcript could be gel purified as described above. No carrier tRNA is added during precipitation of sense RNA.
2. Cyclophilin sense RNA is synthesized similarly, except the plasmid is linearized with *Bam*HI and transcribed with T3 RNA polymerase.
3. To precipitate gel-purified sense RNA, 5 µL of 5 mg/mL glycogen (Ambion) is added, followed by 0.1 vol 3 $M$ NaOAc and 2.5 vol 100% ethanol, and then kept

at –20°C for 30 min. Precipitated RNA is pelleted by centrifugation for 15 min at 4°C in a microfuge. The pellet is washed with 70% ethanol, dried in a Speed Vac for 2–3 min, and dissolved in 10–20 μL of DEPC-H$_2$O.

4. RNA concentration is determined by measuring absorbance at 260 nm in a spectrophotometer.

## 3.3. Solution Hybridization and RNase Digestion

### 3.3.1. Hybridization

1. 5 μL each of labeled NPY (approx 70,000–100,000 cpm) and cyclophilin (approx 15,000–20,000 cpm) cRNA probes are added to a 1.5-mL microfuge tube containing 2–4 μg of medial basal hypothalamus (MBH) RNA (*see* **Note 7**).
2. Carrier tRNA (1.1–1.3 μL) is added to bring the total RNA up to approx 15-20 μg in order to make the RNA pellet easier to visualize.
3. Equal volume (sum of sample, probes, and tRNA volumes) of 5M NH$_4$OAc and 5 vol μL 100% EtOH are added and set at –20°C for 30 min to precipitate RNA.
4. The precipitate is collected by centrifugation for 10 min at 4°C, carefully decanting the supernatant. Residual supernatant liquid is removed by absorption to a Kimwipe.
5. Finally the pellet is vacuum-dried for 2 min in a Speed-Vac and dissolved in 10 μL of hybridization buffer by passing up and down the pipet tip 20 times, followed by vigorous mixing in a vortex mixer.
6. The tube is briefly centrifuged and placed in a 70°C dry bath for 15 min. Thereafter, hybridization is carried on at 45°C overnight (at least 16 h).
7. For negative control, yeast tRNA (approx 15–20 μg) is used with labeled NPY and cyclophilin probes, and hybridization is carried out as described above. The tRNA and probes in this control tube should be completely digested by RNase.

### 3.3.2. RNase Digestion

1. Remove tubes from 45°C water bath, microfuge briefly, and add 290 μL RNase A/T1 digestion buffer to digest unannealed probe. Vortex tubes, and incubate at 32°C for 1 hour.
2. Stop digestion by adding 10 μL 20% SDS and 10 μL proteinase K (10 μg/μL in A/T1 digestion buffer). Vortex briefly, and incubate the tubes at 37°C for 15 min.
3. Add 200 μL phenol and 200 μL chloroform/isoamyl alcohol (24:1) to each tube, vortex for at least 10 s, and microfuge for 3 min at room temperature.
4. Remove the aqueous phase (approx 350 μL), add 10 μg tRNA and 400 μL of 2-propanol, and vortex.
5. Place tubes at –20°C for 30 min, and then microfuge for 10 min at 4°C.
6. Collect the pellet, remove all alcohol (to remove all the salt), and dry the pellet in a Speed-Vac for 2 min (*see* **Note 3**).
7. Dissolve the pellet in 10 μL 80% formamide dye, heat in boiling water bath for 10 min, chill on ice water for 30 s, microfuge briefly, and let stand at room temperature.

8. Prepare tubes with probes (approx 2000 cpm) and labeled Century Marker Plus (approx 5000 cpm; Ambion) in 2 µL volume, add 10 µL of 80% formamide dye to these tubes, and process simultaneously with other samples. By this time gel should be ready to load.

## 3.4. Gel Electrophoresis

1. For one gel, prepare a 50-mL 6% polyacrylamide-8 $M$ urea gel solution, but do not add TEMED.
2. Prepare a vertical gel using 1.5-mm spacers (use the components of the V-16 vertical gel apparatus; Gibco-BRL).
3. Filter gel solution through a Nalgene filter.
4. To seal the glass plates, take 2 mL of gel solution, add 11 µL APS and 18 µL TEMED to it, mix quickly, and pour 1 mL down each side of plate. This will seal the bottom edges of the gel and has been found to prevent leaking of gel solution. However, this procedure must be carried out quickly; otherwise it will be difficult to pour the sealing gel solution because of quick polymerization.
5. Thereafter, add 48 µL TEMED to the remaining gel solution, and swirl. Add gel solution to plates until almost full. Insert comb at an angle to prevent bubbles forming under the wells.
6. Overlay with small amount of water. Allow to polymerize for about 1 h.
7. After polymerization is complete, carefully remove the comb, bottom spacer, and clamps.
8. Remove any unpolymerized gel solution and urea from the wells, and rinse with water (*see* **Note 8**).
9. Place the gel in the apparatus (notched plate to the rear), and hold in place using two clamps at the top. Fill the top chamber with running buffer (1X TBE), check for the leakage, and then fill the bottom chamber. Make sure there are no bubbles being trapped beneath the gel.
10. Thoroughly wash unpolymerized gel solution and urea by using a 5–10-mL syringe and a 25-gage blunt needle (*see* **Note 8**).
11. Prerun the gel for 30 min at approximately 400 V, until gel temperature has reached approx 50°C (*see* **Note 9**).
12. Turn off power, rinse wells once more, and then load samples into wells. Try to keep samples concentrated at the bottom of the wells.
13. Run gel for approximately 2 h at 400 V, or until second dye front has reached the bottom of the gel. Make sure the temperature of the gel stays in between 50 and 55°C throughout the gel electrophoresis (*see* **Note 9**).
14. Turn off the power, and decant the running buffer.
15. Carefully separate gel plates, leaving gel on larger glass plate, and soak gel in gel-soaking solution for 15 min with occasional shaking.
16. Transfer gel to prewetted drying paper, cover the gel with a prewetted cellophane, and dry the gel at 80°C in gel dryer (Bio-Rad) for 90 min.
17. The dried gel is now ready for gel analysis by either autoradiography or phosphorimaging as described below.

## 3.5. Autoradiography and Phosphorimaging

### 3.5.1. Autoradiography

1. Place the dried gel in an autoradiography cassette, and bring into the darkroom. Using the safelight, put one piece of X-ray film in each cassette, and expose with intensifying screen for 16–18 h (or for appropriate period) at –80°C. After exposure, take the cassette to the darkroom, and develop film.
2. Develop in an automatic film developer (Kodak), or manually. For manual film development, open the cassette, and place the film in a tray of diluted developer for 5–10 min (or until developed sufficiently). Wash the film briefly in water, and then put in a tray containing diluted fixer for 5 min. Wash the film extensively with running tap water, and hang dry.
3. Scan the developed film in a densitometer, and quantify protected bands using Molecular Analyst Software (Bio-Rad).

### 3.5.2. PhosphorImaging

1. Erase PhosphorImaging screen for at least 15–20 min before exposure.
2. The dried gel is exposed in a Bio-Rad CS Molecular Imaging Screen for 3–6 h, and the image of each gel is acquired by a GS-525 Molecular Imager (Bio-Rad).
3. The volume analysis of each band obtained from RNase protection assay is performed by Molecular Analyst Software (Bio-Rad). Any PhosphorImaging system can be used. Just follow the protocol or user manual for a specific PhosphorImaginging and data analysis program.

Overall, the protected bands corresponding to NPY and cyclophilin mRNAs are identified according to the expected size. The sizes of the bands are identified according to the molecular weight marker (Century Marker Plus, Ambion).

## 3.6. Data Analysis

PreproNPY mRNA levels are expressed either as a ratio to cyclophilin mRNA or as pg per pg of cyclophilin mRNA. For the first method, the volume of preproNPY mRNA is divided by the volume of cyclophilin mRNA in the same sample. For the second method, the amount of preproNPY mRNA and cyclophilin mRNA are calculated from the standard curve made from respective sense RNAs. Then the amounts of preproNPY mRNA is calculated in terms of pg of cyclophilin mRNA.

## 4. Notes

1. Incomplete RNase A/T1 digestion is usually overcome by using fresh digestion buffer and a new batch of T1 (or check the concentration in the T1 tube, because it varies from batch to batch, and calculate the correct amount needed for digestion). Incomplete RNase A/T1 digestion would be the reason for multiple bands,

only if there are similar bands in the lane with tRNA control. Also check the incubation temperature; it should be 32°C.

2.  RNA degradation could be due to RNase contamination in any of the solutions or any of the tips used for transferring or adding solutions. For this reason, one should wear gloves all the time during RNA extraction and RPA. All solutions must be DEPC treated and autoclaved unless otherwise indicated; also make sure all glass and plastic wares are autoclaved. Finally, since RNase A and T1 are used in the RPA, the investigator must take extra precaution in handling these RNases to avoid unwanted contamination.

3.  We have used Kimwipes (Kimberly-Clark) successfully to dry around the tube in all steps in RNA extraction and RPA without any problem. However, one has to be extremely cautious not to touch the pellet during Kimwiping. Whenever you are ready to use the Kimwipe, wear a new glove, and use an appropriate size Kimwipe such that it has easy access to the tube. A Kimwipe could also be autoclaved if necessary. Alternatively, one can use a drawn-out Pasteur pipet to remove the residual fluid that usually stays in association with the pellet after decanting the solution. In this case, after decanting the solution, remicrofuge the tube for about 5 s, and then use the drawn-out Pasteur pipet.

4.  If it is observed that the probe alone cannot be digested by A/T1, it is most likely because of the presence of DNA template in the probe mixture. If the DNA template is not digested, DNA/RNA hybrid complexes resistant to RNase digestion will be formed, resulting in nonspecific protection. In this situation, one should increase the amount of RQ1 DNase or increase the length of incubation (up to 30–40 min). It is also possible to heat-denature the transcription reaction by heating at 95°C for 2–3 min and quickly chilling in ice water for 30 s, prior to RQ1 DNase treatment at 37°C.

5.  It is particularly important to have a full-length cRNA probe for preproNPY mRNA. Since the length of this probe is around 756 bp (including a 245-bp plasmid sequence), there could be contamination by prematurely terminated shorter length transcripts due to premature termination of transcription. This is usually overcome by gel purification of the probe. Also, it is essential to make sure that there is enough $[^{32}P]UTP$ (approx 6-7 $\mu M$ for an NPY cRNA probe) in the in vitro reaction. Since the cyclophilin probe is approx 117 bp, it does not require gel purification. For NPY sense and antisense RNA, the plasmid could also be linearized with *Sma*I and *Fsp*I, respectively. Restriction enzymes should be chosen according to the cloning site and the vector.

6.  It is advised to use labeled probes on the day of its synthesis to avoid degradation. Also use RNasin (RNA inhibitor) in the solution for dissolving the final probe pellet.

7.  The aforementioned procedures have been successfully used in our studies examining preproNPY mRNA in the hypothalamus. I am confident that this procedure is applicable to preproNPY mRNA measurement in other tissues. However, the amount of total RNA necessary to estimate preproNPY mRNA levels will vary according to the types of tissues and the experimental manipulations.

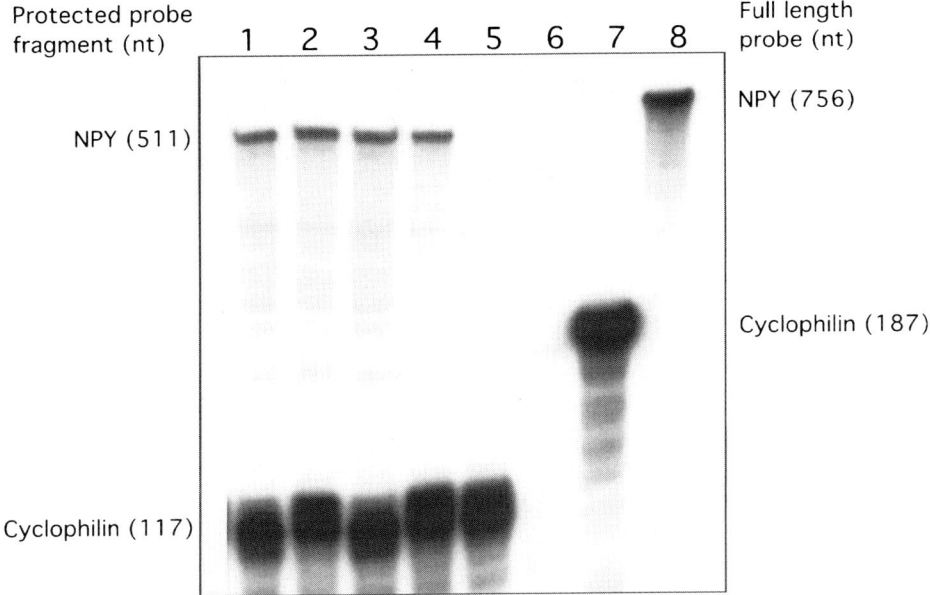

Protected probe fragment (nt)  1 2 3 4 5 6 7 8  Full length probe (nt)

NPY (756)

NPY (511)

Cyclophilin (187)

Cyclophilin (117)

Fig. 1. Autoradiographic bands obtained from RNase protection analysis measuring preproNPY mRNA and cyclophilin mRNA levels in total RNA of the rat MBH. Three micrograms of total RNA were hybridized overnight with $[^{32}P]$UTP-labeled NPY (70,000 cpm) and cyclophilin (15,000 cpm) probes followed by RNase A/T1 digestion for 1 h at 32°C. Protected hybrids were separated on a 6% polyacrylamide-8 $M$ urea denaturing gel. The dried gel was exposed to Fuji Medical X-ray film for 48 h. Lanes 1–4, RNA hybridized with both NPY and cyclophilin probes; lane 5, RNA hybridized with cyclophilin probe alone; lane 6, yeast tRNA hybridized with NPY and cyclophilin probes (note complete digestion of the probes by RNase A/T1); lane 7, 2000 cpm of cyclophilin probe; lane 8, 2000 cpm of NPY probe.

Accordingly, investigators must adjust the amount of total RNA and the time of exposure of the dried gel to the film or PhosphorImager screen after the RPA.

In a typical RPA, only two protected bands were seen in the autoradiograph, one for preproNPY mRNA (upper band) and the other for cyclophilin mRNA (lower band) (*see* **Fig. 1**). If there are many bands in the RPA, then it may be due to several factors such as (1) probe is not of full length; (2) incomplete RNase digestion; (3) presence of DNA in the synthesized labeled probe; (4) degradation of total RNA; and (5) degradation of probe.

8. Gel wells should be washed several times with buffer to remove unpolymerized gel solution and urea in the lane before sample loading. The presence of urea and unpolymerized gel in the wells will interfere with electrophoretic migration through the gel.

9. Put a temperature indicator sticker (Bio-Rad) on the gel plate while running the denaturing gel. Closely monitor the temperature of the gel (approx 50–55°C), because high temperature will distort the gel, and the glass plate may crack. Also check the volume of the buffer in the upper chamber periodically to make sure there is no leakage. Normally leakage should be checked before sample loading by pouring the buffer in the upper chamber first. If there is a leakage during gel running, make sure the clamps are tight; clamps could be added or continue to add buffer periodically into the upper chamber, but always turn off the power supply before you add buffer.

## Acknowledgments

This work was supported by NIH grant RO1 AG-10868.

## References:

1. Krause, J. E., Cremins, J. D., Carter, M. S., Brown, E. R., and MacDonald, M. R. (1989) Solution hybridization-nuclease protection assays for sensitive detection of differentially spliced substance P- and neurokinin A-encoding messenger ribonucleic acids, in *Methods in Enzymology*, vol 168 (Conn, P. M., ed.), Academic, San Diego, pp. 634–652.
2. Blum, M. (1989) Regulation of neuroendocrine peptide gene expression, in *Methods in Enzymology*, vol 168 (Conn, P. M., ed.), Academic, San Diego, pp. 618–633.
3. Chomczynski, P. and Sacchi, N. (1987) Single step method of RNA isolation by acid guanidium thiocyanate-phenol-chloroform extraction. *Anal. Biochem.* **162,** 156–159.
4. Sambrook, J., Fritch, E. F., and Maniatis, T. (1989) *Molecular Cloning: A Laboratory Manual*, 2nd ed. Cold Spring Harbor Laboratory, Cold Spring Harbor, NY.

# 18

# Radioligand Binding Studies

*Pharmacological Profiles of Cloned Y-Receptor Subtypes*

**Karen E. McCrea and Herbert Herzog**

## 1. Introduction

Radioligand binding is an extremely powerful technique that can provide detailed information about receptor-ligand interactions both in vitro and in vivo. Several types of binding assay can be performed, including studies of the kinetics of association or dissociation of the radiolabeled ligand, and equilibrium binding experiments such as saturation and competition binding assays. Data obtained from such studies generate accurate information regarding receptor number, ligand affinity, the existence of receptor subtypes, and allosteric interactions between binding sites and/or receptors. Generally receptor-ligand interactions are studied using membrane preparations since these give highly reproducible results. However, the technique can also be adapted to other preparations such as whole cells, tissue slices, solubilized receptors, and even whole animals. The techniques used and problems associated with radioligand binding studies are reviewed in **ref. 1**.

Radioligand binding has been a particularly useful tool in demonstrating the existence of various neuropeptide (NPY) receptor (Y receptor) subtypes. Originally, Y receptors were subdivided into two classes based on pharmacological properties. Certain C-terminal fragments of NPY, e.g., [13–36]NPY were, like NPY, capable of inhibiting prejunctional twitch responses in rat vas deferens but had no effect on other NPY-mediated responses such as vasoconstriction in guinea pig iliac vein (**2**). It was therefore proposed that two subclasses of Y receptor existed: postjunctional Y1 receptors, which are activated by holopeptides, and prejunctional Y2 receptors, which are activated by both holopeptides and C-terminal fragments (**2**).

From: *Methods in Molecular Biology, vol. 153: Neuropeptide Y Protocols*
Edited by: A. Balasubramaniam © Humana Press Inc., Totowa, NJ

Since this initial classification, a large number of Y-receptor subtypes have been identified using molecular cloning techniques, including Y1, Y2, Y4, Y5, and Y6 *(3)*. Pharmacological investigations have led to the suggestion that other putative Y-receptor subtypes may exist, such as Y3, a receptor that appears to have a higher affinity for NPY than peptide YY (PYY), and a "PYY-preferring receptor" in the periphery *(3)*.

Unfortunately, the ability to clone multiple Y-receptor subtypes has not been matched by the development of selective agonists and antagonists. This has led to difficulty in assigning particular functions for Y-receptor subtypes in vivo. Furthermore, various laboratories use a range of radiolabels, competing ligands from diverse species, different buffer components, assay temperatures, and incubation times to study Y-receptor pharmacology in vitro. This has led to conflicting results concerning peptide affinities for a particular Y-receptor subtype. For example, the order of affinity of a range of ligands for the mouse Y6 receptor alters depending on the buffer or radiolabel employed *(4,5)*.

We have conducted radioligand binding studies using a system that aims to keep these factors constant in an attempt to compare ligand affinity for a particular subtype. We have subcloned each of the four human Y receptors into the same expression vector, pcDNA3, and transfected them into human embryonic kidney (HEK 293) cells. Following the establishment of stable clonal cell lines, the ligand binding properties of a range of NPY peptides and associated peptide fragments have been studied using an $^{125}$I-labeled version of the most abundant natural ligand, NPY. In addition, all assays are performed using the same buffer system, incubation temperature, and incubation time to provide a valid comparison of ligand affinities between Y-receptor subtypes.

## 2. Materials

1. Crude membrane preparations: prepared as indicated in **Subheading 3.1.**
2. Unless otherwise stated, all reagents were obtained from Sigma, St. Louis, MO.
3. Dulbecco's phosphate-buffered saline (PBS): 138 m$M$ NaCl, 8.1 m$M$ Na$_2$HPO$_4$, 2.5 m$M$ KCl, 1.2 m$M$ KH$_2$PO$_4$, pH 7.4 (GIBCO, Life Technologies, Gaithersburg, MD, #14190-144).
4. Homogenization buffer: 50 m$M$ Tris-HCl, 10 m$M$ NaCl, 5 m$M$ MgCl$_2$, 2.5 m$M$ CaCl$_2$, pH 7.4, stored at 4°C and supplemented with 1 mg/mL bacitracin (250,000 U; Calbiochem-Novabiochem., La Jolla, CA, #1951) prior to use (*see* **Note 1**).
5. Assay buffer: same as homogenization buffer except supplemented with 1 mg/mL bacitracin and 0.1% (w/v) bovine serum albumin (BSA; Fraction V, Sigma #A-9418) prior to use. Both bacitracin and BSA should be weighed out freshly on the day of use.
6. Neuropeptides: diluted to a stock concentration of 250 µ$M$ using 0.01 $M$ acetic acid and stored in small aliquots at –80°C to avoid freeze-thawing. Serial dilutions are made in assay buffer just prior to use and stored on ice (do not reuse).

The peptides employed for routine investigations in this laboratory include NPY, PYY, pancreatic polypeptide (PP), [Leu$^{31}$Pro$^{34}$]NPY, [2–36]NPY, [3–36]NPY, and [13–36]PYY. All peptides were obtained from Auspep (Parkville, Australia).

7. [$^{125}$I]porcineNPY (Bolton Hunter labeled, i.e., iodine molecule is conjugated to the ε-amino group of Lys$^4$ of NPY) available from NEN DuPont (Boston, MA, #NEX222, specific activity 2200 Ci/mmol). Aliquotes stored at –80°C (*see* **Note 2**).
8. Horse serum (CSL Biosciences, Parkville, Australia, #09532301): 50-mL aliquots stored at –20°C (*see* **Note 1**).
9. GF/C glass microfiber filters (Whatman, Maidstone, UK, #18 22915). Sheets (46 × 57 cm) of glass fiber filters cut to size (35 × 6 cm) just prior to use.
10. Cell harvester filtering apparatus (Brandel Instruments, Gaithersburg, MD).
11. 50% (w/v) aqueous polyethyleneimine (PEI; Sigma #P-3143) diluted to 0.3% (v/v) with water.
12. Cell culture flasks (150 cm$^2$) (#430823) and cell scrapers (#3011) were obtained from Corning Costar (Cambridge, MA).

## 3. Methods
### 3.1. Preparation of Membranes

1. Grow the appropriate cell line to confluency in 8 × T150 cell culture flasks.
2. Wash confluent monolayers twice with 10 mL ice-cold PBS (pH 7.4).
3. Remove confluent monolayers sequentially from each flask in 10 mL ice-cold homogenization buffer with a cell scraper.
4. Disrupt cells by hand with a 15-mL ground glass homogenizer (Wheaton, USA) using 15 strokes. All homogenization procedures must be performed on ice.
5. Rinse each flask sequentially with a further 10 mL homogenization buffer, and homogenize (15 strokes).
6. Pool homogenates into 30-mL polyethylene tubes, and centrifuge at 32,000$g$ for 15 min at 4°C.
7. Resuspend the resulting pellet in 8 mL ice-cold homogenization buffer and rehomogenize (15 strokes).
8. Centrifuge again at 32,000$g$ for 15 min at 4°C.
9. Resuspend the final pellet to 10–20 mg protein/mL in ice-cold homogenization buffer, and flash freeze in liquid N$_2$.
10. Store samples at –80°C until required. Samples can be stored for at least 6 months without alteration of binding properties.
11. Measure protein concentration using standard procedures (e.g., Bradford method using Bio-Rad Reagent, with BSA as standard).

### 3.2. Radioligand Binding Assays
#### 3.2.1. Saturation Binding Assay

1. Final volume for each assay is 250 μL. For filtration and centrifugation assays, incubations should be performed in polypropylene culture tubes or 1.5-mL Eppendorf tubes, respectively.

2. Add 50-μL assay buffer to all tubes.
3. Add 50 μL NPY (1 μ*M* ) to tubes to define nonspecific binding (NSB; see **Note 3**) or add 50 μL assay buffer to define total binding (TB).
4. Add 50 μL [$^{125}$I]NPY (0.001–1 p*M*) to all tubes.
5. Initiate the incubation by the addition of cell membrane suspensions (100 μL). Cell membranes should have been previously diluted to yield an optimal protein concentration so that TB <10% of total counts (TC; see **Note 4**). Mix thoroughly by vortexing.
6. Incubate at room temperature for 2 h.
7. Separate bound from free [$^{125}$I]NPY by filtration (*see* **Subheading 3.2.3.**) or centrifugation (*see* **Subheading 3.2.4.** and **Note 5**).

## 3.2.2. Competition Binding Assay

1. Final volume for each assay is 250 μL.
2. Add 50 μL assay buffer to all tubes.
3. Add 50 μL NPY (1 μ*M* ) to NSB tubes (*see* **Note 3**) or add 50 μL assay buffer to TB tubes. Add 50 mL "cold" ligand (e.g., serial dilutions from $10^{-6}$ to $10^{-12} M$) to all remaining tubes.
4. Add 50 μL [$^{125}$I]NPY (e.g., 50 p*M*) to all tubes.
5. Initiate the incubation by the addition of cell membrane suspension (100 μL). Cell membranes should have previously been diluted to yield an optimal protein concentration so that TB <10% TC. Mix thoroughly by vortexing.
6. Incubate at room temperature for 2 h.
7. Separate bound from free [$^{125}$I]NPY by filtration (*see* **Subheading 3.2.3.**) or centrifugation (*see* **Subheading 3.2.4.**).

## 3.2.3. Separation of Bound from Free [$^{125}$I]NPY by Filtration

1. Presoak Whatman GF/C filters in 0.3% PEI for at least 2 h.
2. Terminate incubations by rapid filtration over filters using a cell harvester filtering apparatus.
3. Wash filters (3 × 4 mL) with ice-cold 50 m*M* Tris-HCl (pH 7.4; *see* **Note 6**).
4. Punch filters into plastic vials, and determine radioactivity immediately using a γ-counter.

## 3.2.4 Separation of Bound from Free [$^{125}$I]NPY by Centrifugation

1. Layer each sample over 0.5 mL horse serum (precooled to 4°C; *see* **Note 6**).
2. Centrifuge samples (13,000*g*) for 4 min.
3. Remove supernatant solution and horse serum by aspiration.
4. Cut off the tip of the Eppendorf tube containing the pellet (bottom 5 mm; we find that dog nail clippers are very useful in this procedure), and place in a plastic tube.
5. Determine radioactivity immediately using a γ-counter.

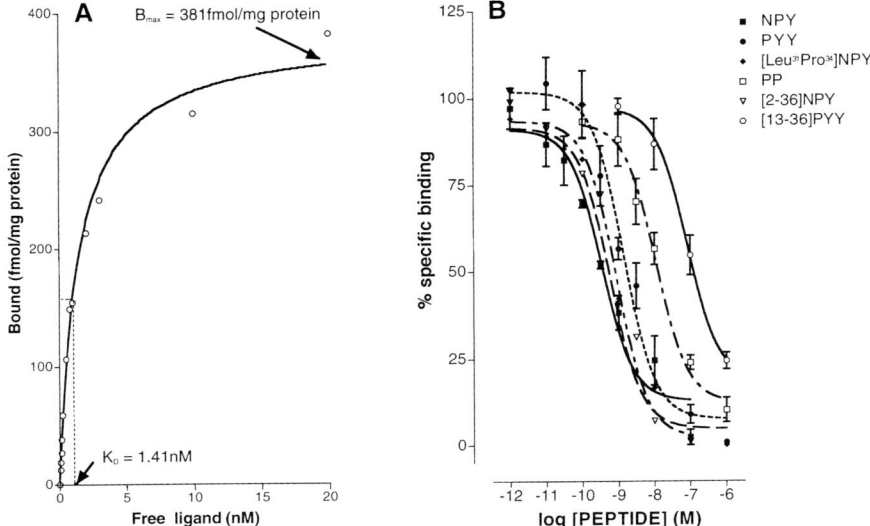

Fig. 1. (**A**) [$^{125}$I]NPY saturation curve showing specific binding as a function of free ligand concentration at the human Y1 receptor. (**B**) Competition curves for a range of peptide analogs at the human Y5 receptor in the presence of a fixed concentration of [$^{125}$I]NPY (50 p*M*).

### 3.3. Data Analysis

Data are represented in graphical form in terms of the amount of radiolabel bound (y-axis) versus (1) radiolabel concentration in saturation experiments (**Fig. 1A**) or (2) displacing ligand concentration ($\log_{10}$ units) in competitive binding assays (**Fig. 1B**). The graphics package Prism (GraphPad Software, San Diego, CA) is excellent for displaying such data.

Often, linear transformations of binding curves are used to exhibit binding complexities and provide estimations of binding parameters (*see* **Note 6**). For saturation experiments, Scatchard plots are generally employed. Data are represented as bound/free radiolabel (*y*-axis) versus bound radiolabel (*x*-axis; **Fig. 2**). If data points form a straight line, it can be assumed that the radiolabel binds to a single population of binding sites. The slope of the line represents -1/$K_D$, and the x-axis intercept represents $B_{max}$ (**Fig. 2**). If a curved plot is obtained, this indicates the existence of binding sites with different affinities for the radiolabel or that positive or negative cooperativity is occurring (*see* **Notes 7** and **8**). Scatchard plots can be unreliable unless data points are extremely accurate, and linear regression should not be applied to such plots since $K_D$ and $B_{max}$ values will invariably be overestimated (*6; see* **Note 9**).

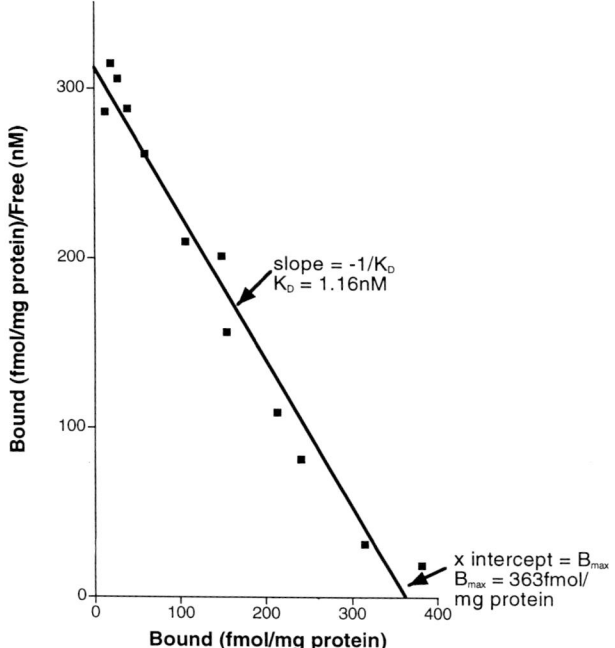

Fig. 2. Scatchard plot of specific binding of [$^{125}$I]NPY to the human Y1 receptor, showing estimates of $K_D$ and $B_{max}$.

Instead of using graphical methods for estimations of $B_{max}$ and $K_D$ values, data should be analyzed using a computerized statistical package such as LIGAND (Biosoft, Cambridge, UK) where curve fitting by weighted nonlinear least-squares is employed (7).

It is possible to determine equilibrium dissociation constants ($K_i$) for unlabeled ligands from competition studies. $K_i$ values can be calculated from IC$_{50}$ values (concentration of "cold" ligand that reduces specific binding by 50%) obtained by nonlinear least-squares analysis of the displacement curve (8).

$$K_i = IC_{50} / (1 + [L]/K_D)$$

IC$_{50}$ values in displacement curves only give a direct estimation of $K_i$ if [L] (the concentration of radiolabel employed) is very low relative to the $K_D$ (*see* **Note 10**).

## 4. Notes

1. It is very important to batch test new stocks of bacitracin and horse serum prior to use. On several occasions a reduction or complete loss of specific binding has

been observed when new batches were employed. If other protease inhibitors such as leupeptin or aprotinin are to be included in assay buffers, batch testing is also recommended.

2. Breakdown of [$^{125}$I]NPY during storage is minimized by following the manufacturer's instructions. Generally each stock solution is aliquoted into smaller volumes to avoid freeze-thawing procedures. Since [$^{125}$I]radiolabels decay rapidly (half-life of 60 days), it is essential to correct for decay on the day of use. To ensure that the correct concentration of radiolabel is being used in each experiment, it is also important to check the calculation by counting a small sample of the diluted radiolabel stock. [$^{125}$I]NPY is stable and usable for at least 6 weeks after manufacture, although NSB increases to some extent during this time. Similar precautions should be taken when using [$^{125}$I]PYY or [$^{125}$I]PP.

3. Since [$^{125}$I]NPY binds to sites other than Y receptors on membrane preparations, it is essential to separate nonspecific binding from specific binding. Ideally, a competing ligand that is chemically distinct from [$^{125}$I]NPY and employed at $100–1000 \times$ its $K_D$ should be used to define NSB. However, a high concentration of unlabeled NPY can give a reliable estimate of NSB since all competing ligands that exhibit high affinity for a particular Y-receptor subtype produce identical maximal inhibition of [$^{125}$I]NPY binding in competition studies.

4. It is important to ensure that depletion of radiolabel and/or competing ligands does not occur, i.e., free radiolabel/ligand concentration must not be significantly different from the total concentration added. Ligand depletion occurs when the protein concentration of the membrane preparation is too high or when loss of ligand occurs during dilution or storage, or due to the ligand sticking to tubes or pipet tips. Ligand depletion will cause steepening of saturation binding curves or a significant rightward shift in competition binding curves. To minimize the problem, it is essential to keep the receptor concentration as low as possible, while maintaining adequate specific binding, so that depletion of radiolabel is <10% and therefore of no real significance. In centrifugation assays, it is also possible to measure the free concentration of radiolabel readily in supernatant fractions and compare this with "total" counts. Unfortunately this cannot be readily done in filtration assays. In principle, the wash solution can be carefully collected and used to measure unbound ligand. Alternatively, a centrifugation assay can be performed in parallel.

5. [$^{125}$I]NPY is a "sticky" radiolabel, which is probably one reason why it is not generally used to study Y-receptor pharmacological profiles in many laboratories. As such, its use leads to high levels of NSB in filtration assays. High levels of NSB can be minimized by using a low concentration of radiolabel, and a minimal concentration of membrane protein and by presoaking glass fiber filters in 0.3% PEI, which inhibits NSB of [$^{125}$I]radiolabel. Under such conditions, NSB is kept to a low level when membrane preparations expressing high levels of receptor are employed. Clonal cell lines that exhibit a much lower density of receptor sites require substantially more protein/tube to achieve measurable specific binding, thus raising NSB to unacceptable levels. Under these circumstances we have

found that separation of bound and free radiolabel is better achieved using a centrifugation assay in which samples are separated through horse serum. In parallel experiments, NSB levels are lower using this technique than those determined with filtration assays. [$^{125}$I]PYY and [$^{125}$I]PP are not as "sticky" as [$^{125}$I]NPY, and filtration techniques are ideally suited to separate bound from free radiolabel using those radioligands.

6. Dissociation of specifically bound radiolabel from Y receptors is hindered at low temperatures. Thus, during separation of bound and free ligand, dissociation can be minimized by washing filters with ice-cold buffer (filtration assay) or layering samples over horse serum at 4°C (centrifugation assay).

7. Addition of an iodine molecule to a ligand can produce a derivative that exhibits markedly different chemical properties when compared with the native peptide. Therefore, the $K_D$ calculated for [$^{125}$I]NPY in saturation binding studies may not be the same as the $K_i$ for "cold" NPY obtained from competition studies at a particular Y-receptor subtype.

8. The experiments described here employ [$^{125}$I]NPY as radiolabel. The human Y4 receptor exhibits a particularly low affinity for [$^{125}$I]NPY. [$^{125}$I]PP is generally employed to determine the pharmacological profile for this receptor subtype .

9. The use of [$^{125}$I]NPY, an agonist, has a number of associated problems not encountered when a radiolabeled antagonist is employed in binding studies. The major drawback is that agonists binding to receptors that couple to G-proteins can interact with high and low affinities. The choice of radioligand, [$^{125}$I]NPY or [$^{125}$I]PYY, in competition studies can produce completely different binding profiles for a range of ligands.

10. It should be noted that when the radioligand binding properties for novel Y-receptor subtypes are being investigated, it may be necessary to modify certain assay conditions. For example, the novel receptor may exhibit a low affinity for NPY. Therefore it would be prudent to measure binding parameters using a range of radiolabeled peptides. As was observed when the third galanin receptor (GALR3) was cloned, specific binding was only achieved using specific assay buffer components *(9)*. Thus, it is important to realize that although radioligand binding is a very simple technique, many factors must be taken into account to achieve optimal assay conditions.

## References

1. Hulme, E. C. and Birdsall, N. J. (1992) Strategy and tactics in receptor binding studies, in *Receptor-Ligand Interactions. A Practical Approach* (Hulme, E. C., ed.), IRL Press at Oxford University Press, Oxford, pp. 63–177.

2. Wahlestedt, C., Yanaihara, N., and Hakanson, R. (1986) Evidence for different pre and post-junctional receptors for neuropeptide Y and related peptides. *Regul. Pept.* **13,** 307–318.

3. Blomqvist, A. G. and Herzog, H. (1997) Y-receptor subtypes—how many more? *Trends Neurosci.* **20,** 294–298.

4. Weinberg, D. H., Sirinathsinghji, D. J. S., Tan, C. P., Shiao, L.-L., Morin, N., Rigby, M. R., et al. (1996) Cloning and expression of a novel neuropeptide Y receptor. *J. Biol. Chem.* **271,** 16,435–16,438.

5. Gregor, P., Feng, Y., DeCarr, L. B., Cornfield, L. J., and McCaleb, M. L. (1996) Molecular characterization of a second mouse pancreatic polypeptide receptor and its inactivated human homologue. *J. Biol. Chem.* **271,** 27,776–27,781.

6. Burgisser, E. (1984) Radioligand-receptor binding studies: what's wrong with the Scatchard analysis? *Trends Pharmacol. Sci.* **5,** 142–144.

7. Munson, P. J. and Rodbard, D. (1980) LIGAND: a versatile computerized approach for the characterization of ligand binding systems. *Anal. Biochem.* **107,** 220–239.

8. Cheng, Y. C. and Prusoff, W. H. (1973) Relationship between the inhibition constant Ki and the concentration of inhibitor which causes 50% inhibition (I50) of an enzymatic reaction. *Biochem. Pharmacol.* **22,** 3099–3108.

9. Wang, S., He, C., Hashemi, T. and Bayne, M. (1997) Cloning and expressional characterization of a novel galanin receptor. *J. Biol. Chem.* **272,** 31,949–31,952.

10. Fahti, Z., Battaglino, P. M., Iben, L. G., Li, H., Baker, E., Zhang, D., et al. (1998) Molecular characterization, pharmacological properties and chromosomal localization of the human GALR2 galanin receptor. *Mol. Brain Res.* **58,** 156–169.

# Index

From: *Methods in Molecular Biology, vol. 153, Neuropeptide Y Protocols*
Edited by: A. Balasubramaniam © Humana Press Inc., Totowa, NJ